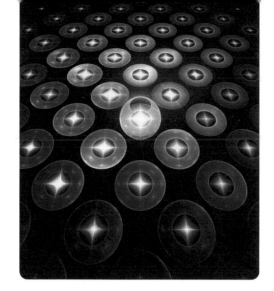

WJEC
Mathematics
for A2 Level – Applied

Stephen Doyle

Illuminate Publishing

Published in 2019 by Illuminate Publishing Limited, an imprint of Hodder Education, an Hachette UK Company, Carmelite House, 50 Victoria Embankment, London EC4Y 0DZ

Orders: please contact Hachette UK Distribution, Hely Hutchinson Centre, Milton Road, Didcot, Oxfordshire, OX11 7HH. Telephone: +44 (0)1235 827827. Email: education@hachette.co.uk. Lines are open from 9 a.m. to 5 p.m., Monday to Friday. You can also order through our website: www.hoddereducation.co.uk

British Library Cataloguing in Publication Data

A catalogue record for this book is available from the British Library

ISBN 978-1-911208-55-6

Printed by Ashford Press in the UK

Impression 3
Year 2024

This material has been endorsed by WJEC and offers high quality support for the delivery of WJEC qualifications. While this material has been through a WJEC quality assurance process, all responsibility for the content remains with the publisher.

Hachette UK's policy is to use papers that are natural, renewable and recyclable products and made from wood grown in well-managed forests and other controlled sources. The logging and manufacturing processes are expected to conform to the environment regulations of the country of origin.

Editor: Geoff Tuttle
Cover design: Neil Sutton
Text design and layout: GreenGate Publishing Services, Tonbridge, Kent

Photo credits

Cover: Klavdiya Krinichnaya/Shutterstock; **p9** Dark Moon Pictures/Shutterstock; **p32** Russell Parry; **p60** iMoved Studio/Shutterstock; **p83** Elana Erasmus/ Shutterstock; **p93** Serhii Moiseiev /Shutterstock; **p103** drpnncpptak/Shutterstock; **p119** Alexey Broslavets/Shutterstock; **p139** Aspen Photo/Shutterstock; **p153** Martin Lisner/Shutterstock.

Acknowledgements

The author and publisher wish to thank Sam Hartburn and Siok Barham for their careful attention when reviewing this book.

Contents

Contents

How to use this book

The contents of this study and revision guide are designed to guide you through to success in the Applied Mathematics component of the WJEC Mathematics for A2 level: Applied examination. It has been written by an experienced author and teacher and edited by a senior subject expert. This book has been written specifically for the WJEC A2 course you are taking and includes everything you need to know to perform well in your exams.

Knowledge and Understanding

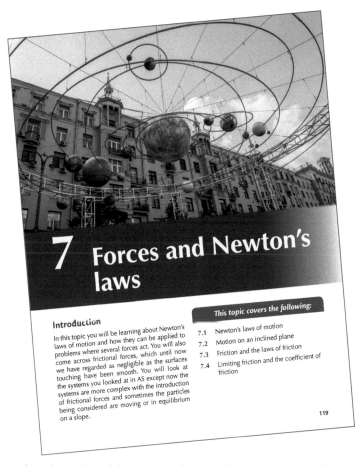

Topics start with a short list of the material covered in the topic and each topic will give the underpinning knowledge and skills you need to perform well in your exams.

The knowledge section is kept fairly short leaving plenty of space for detailed explanation of examples. Pointers will be given to the theory, examples and questions that will help you understand the thinking behind the steps. You will also be given detailed advice when it is needed.

The following features are included in the knowledge and understanding sections:

- **Grade boosts** – are tips to help you achieve your best grade by avoiding certain pitfalls which can let students down.

- **Step by steps** – are included to help you answer questions that do not guide you bit by bit towards the final answer (called unstructured questions). In the past you would be guided to the final answer by the question being structured. For example, there may have been parts (a), (b), (c) and (d). Now you can get questions which ask you to go to the answer to part (d) on your own. You have to work out for yourself the steps (a), (b) and (c) you would need to take to arrive at the final answer. The 'step by steps' help teach you to look carefully at the question to analyse what steps need to be completed in order to arrive at the answer.

- **Active learning** – are short tasks which you carry out on your own which aid understanding of a topic or help with revision.

- **Summaries** – are provided for each topic and present the formulae and the main points in a topic. They can be used for quick reference or help with your revision.

Exam Practice and Technique

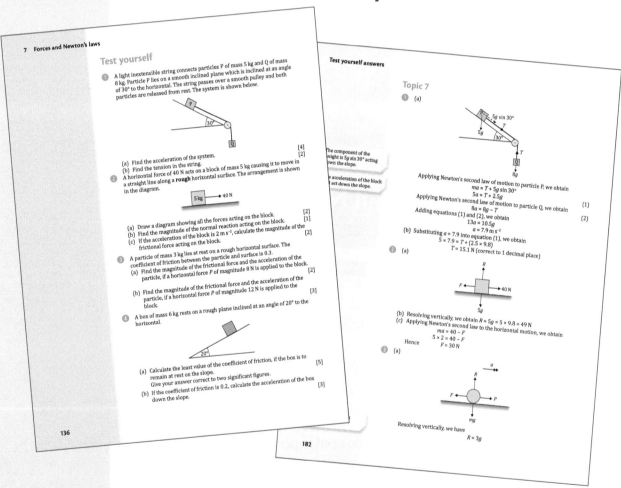

Helping you understand how to answer examination questions lies at the heart of this book. This means that we have included questions throughout the book that will build up your skills and knowledge until you are at a stage to answer full exam questions on your own. Examples are included; some of which are full examination style questions. These are annotated with Pointers and general advice about the knowledge, skills and techniques needed to answer them.

There is a Test yourself section where you are encouraged to answer questions on the topic and then compare your answers with the ones given at the back of the book. There are many examination-standard questions in each test yourself that provides questions with commentary so you can see how the question should be answered.

You should, of course, work through complete examination papers as part of your revision process.

We advise that you look at the WJEC website www.wjec.co.uk where you can download materials such as the specification and past papers to help you with your studies. From this website you will be able to download the formula booklet that you will use in your examinations. You will also find specimen papers and mark schemes on the site.

Good luck with your revision.

Stephen Doyle

WJEC Mathematics For A2 Level
Pure & Applied Practice Tests

There is another book which can be used alongside this book. This book provides extra testing on each topic and provides some exam style test papers for you to try. I would strongly recommend that you get a copy of this and use it alongside this book.

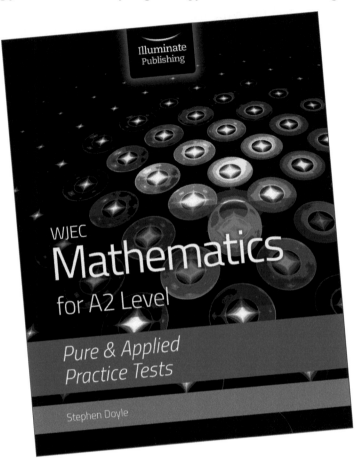

1 Probability

Introduction

You were introduced to probability in the AS course but before starting this topic you should take a look back at the AS Applied book at Topic 3 Probability on pages 55 to 68.

In the AS, you looked at the probability of independent events, which meant that the probability of a first event occurring did not influence the probability of a second event occurring. Such events are called independent events. In many cases the probability of the second event occurring is dependent on whether or not the first event occurs and such events are called dependent events and the probability of the events occurring is called conditional probability.

This topic looks at the probability of dependent events and the various methods of finding their probability.

This topic covers the following:

1.1 Using tree diagrams for conditional probability

1.2 The conditional probability formula

1.3 Use of Venn diagrams for conditional probability

1.4 Use of two-way tables for conditional probability

1.5 Modelling with probability

1.1 Using tree diagrams for conditional probability

You came across tree diagrams in your GCSE studies. Tree diagrams are diagrams used to represent the probabilities of combined events. Each path (which is a branch of the tree) corresponds to a certain sequence of events. By multiplying the probabilities of the separate events along the path you can work out the probability of a particular sequence of events. If the required probability involves several paths, the probabilities of each path are found and then added together to give the required probability.

Tree diagrams for conditional probability

Conditional probability is the probability of an event occurring given that another event has occurred.

Suppose you have a bag containing 6 red and 4 black balls and one ball is picked at random and its colour noted and kept out of the bag. Another ball is picked at random and its colour noted. We can draw the following tree diagram to show the probabilities:

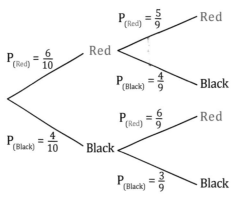

Notice that on the second pick the probability has changed depending on what colour ball was removed in the first pick.

If the first ball had been put back into the bag ready for picking the second ball, the probabilities would have been the same as the first pick.

In the situation shown by the tree diagram, the probability of obtaining each colour has changed depending on what colour ball was removed in the first pick. Hence, this is conditional probability.

For example, if you wanted the probability that the first ball was black and the second ball was red you multiply the probabilities along the path like this

$$P(\text{black and then red}) = P(\text{black}) \times P(\text{red given first ball was black})$$

We can write this using the following shorthand notation

$$P(B \cap R) = P(B) \times P(R|B)$$

$$= \frac{4}{10} \times \frac{6}{9}$$

$$= \frac{4}{15}$$

Note that P($B \cap R$) is the probability of obtaining a black and then a red ball, P($R|B$), is the probability of obtaining a red ball given that a black had been obtained from the first pick.

Suppose there are two events A and B. The probability of B occurring given that A has already occurred is written as P($B|A$). For short, we can say that P($B|A$) is the probability of B given A.

Conditional probabilities can be shown using the following tree diagram:

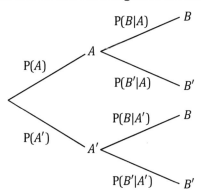

The probability of A and B both occurring is given by

$$P(A \cap B) = P(A) \times P(B|A)$$

This is an important law and is called the multiplication law for dependent events.

By using the above tree diagram, we have the following results:

$$P(A \cap B') = P(A) \times P(B'|A)$$

$$P(A' \cap B) = P(A') \times P(B|A')$$

$$P(A' \cap B') = P(A') \times P(B'|A')$$

Notice the way any of these formulae can be produced using the formula given in the formula booklet (i.e. $P(A \cap B) = P(A) \times P(B|A)$ simply by changing A to A' or B to B' or both A and B to A' and B').

We can also have

$$P(A \cap B) = P(B) \times P(A|B)$$

1.2 The conditional probability formula

The multiplication law for dependent events is

$$P(A \cap B) = P(A)\, P(B|A)$$

and also,

$$P(A \cap B) = P(B)\, P(A|B)$$

Examples

1 Two events A and B occur such that $P(A') = 0.6$, $P(B|A) = 0.7$, $P(B|A') = 0.4$.

(a) Copy and complete the tree diagram shown below to illustrate the above information.

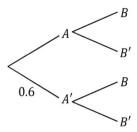

(b) Using your tree diagram, find:

    (i)   P($A \cap B$)

    (ii) P($A' \cap B'$)

    (iii) P($B'|A$)

Answer

1 (a)

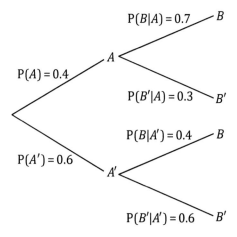

BOOST

Grade ⇧⇧⇧⇧

It is a good idea to write the probability in full rather than just the number on each of the branches, e.g. rather than write 0.4 on the branch for *A*, you would write P(*A*) = 0.4. It takes a bit longer but is helpful when you have to work out probabilities.

(b) (i)   P($A \cap B$) = P(A) × P($B|A$) = 0.4 × 0.7 = 0.28

     (ii) P($A' \cap B'$) = P(A') × P($B'|A'$) = 0.6 × 0.6 = 0.36

     (iii) P($B'|A$) = 0.3

---

2 Two events *A* and *B* occur such that P($B|A$) = 0.55, P($B|A'$) = 0.5 and P(A) = 0.65

  (a) Draw a tree diagram to represent this information.

  (b) Find the following probabilities

    (i)   P($A \cap B$)

    (ii) P(B)

    (iii) P($A|B$)

Answer

2 (a)

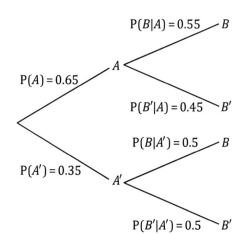

(b) (i) $P(A \cap B) = P(A) \times P(B|A)$

$= 0.65 \times 0.55$

$= 0.3575$

(ii) $P(B) = P(A) \times P(B|A) + P(A') \times P(B|A')$

$= 0.65 \times 0.55 + 0.35 \times 0.5$

$= 0.5325$

(iii) $P(A|B) = \dfrac{P(A \cap B)}{P(B)}$

$= \dfrac{0.3575}{0.5325}$

$= 0.6714$

> Note that there are two paths giving the probability that B occurs.

---

3 When Ahmed travels to college by car, the probability of him being late is 0.2. If he does not travel by car, the probability of him being late is 0.3. The probability of him travelling by car is 0.4.

Find the probability that Ahmed is late.

. . . .

Answer

3   Let C = event that Ahmed travels by car.

L = event that Ahmed is late.

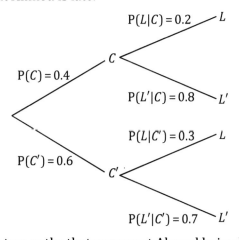

$P(L|C) = 0.2$ L

$P(C) = 0.4$ C

$P(L'|C) = 0.8$ L'

$P(L|C') = 0.3$ L

$P(C') = 0.6$ C'

$P(L'|C') = 0.7$ L'

> **BOOST**
>
> **Grade** ⇧⇧⇧⇧
>
> It is easier to define the meaning of the letters you are using first and then only use these letters on the tree diagram. The probabilities will then be expressed in the same way as that used in any of the formulae.

Notice there are two paths that represent Ahmed being late.

$P(L) = P(C \cap L) + P(C' \cap L)$

$= P(C) \times P(L|C) + P(C') \times P(L|C')$

$= 0.4 \times 0.2 + 0.6 \times 0.3$

$= 0.26$

4   Jack is taking part in a quiz programme. For each question in the quiz, four alternative answers are given, only one of which is correct. Jack has probability 0.6 of knowing the correct answer to a question, and when he does not know the correct answer, he chooses one of the four answers at random.

(a) Calculate the probability that Jack gives the correct answer to a question.

(b) Given that Jack gave the correct answer to a question, find the probability that he knew the correct answer.

Answer

> It is always advisable to draw a tree diagram unless the number of branches makes it too difficult to draw.

4   (a) Let K = event knowing the correct answer

A = event giving the correct answer

> Notice that once Jack knows the correct answer, the probability then of him giving the correct answer is 1.

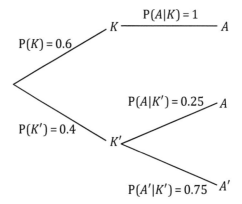

> There are two paths to consider on the tree diagram.

P(giving correct answer) = $(0.6 \times 1) + (0.4 \times 0.25)$

= 0.7

(b) P(knew correct answer|gave the correct answer)

$= \dfrac{\text{P(knew the correct answer)}}{\text{P(gave correct answer)}} = \dfrac{0.6}{0.7} = 0.857\,(\text{correct to 3 d.p.})$

Determining whether two events are independent or not

Suppose we have two events A and B and we want to find out if they are independent events (i.e. the occurrence of one event has no influence on the other event occurring) or dependent events, we can use the method shown here.

If the two events are independent, then the multiplication law for independent events is true, so:

$$P(A \cap B) = P(A) \times P(B)$$

The generalised addition law, shown below, can be used for dependent or independent events.

$$P(A \cap B) = P(A) + P(B) - P(A \cup B)$$

Hence if you found the value of $P(A \cap B)$ using the generalised addition law and the value was not the same as that found using the multiplication law, then the events A and B are dependent.

If the values for P($A \cap B$) obtained using both formulae are the same, then events A and B must be independent.

The following examples show this method.

Examples

1 If two events A and B occur such that P(A) = 0.4 and P(B) = 0.8 and P($A \cap B$) = 0.32, determine whether or not these two events are independent.

· ·

Answer

1 If the events are independent, then P($A \cap B$) = P(A) × P(B).

We can enter the values for the probabilities of A and B respectively and then compare the value of P($A \cap B$) calculated with the value in the question. If they are the same, then A and B are independent events.

$$P(A \cap B) = P(A) \times P(B)$$
$$= 0.4 \times 0.8$$
$$= 0.32$$

This value is the same as the value in the question which means A and B are independent events.

---

2 Two events A and B are such that P(A) = 0.2, P(B) = 0.6 and P($A \cup B$) = 0.4.

Decide whether events A and B are dependent or independent events.

· ·

Answer

2 The generalised addition law that links the probability of the intersection ($A \cap B$) with the probability of the union ($A \cup B$) can be used here.

$$P(A \cap B) = P(A) + P(B) - P(A \cup B)$$
$$= 0.2 + 0.6 - 0.4$$
$$= 0.4$$

> P($A \cap B$) is the probability of A and B occurring. The generalised addition law is included in the formula booklet.

If the events A and B are independent, the probability of both A and B occurring is found by multiplying P(A) and P(B) together.

So if the events are independent $P(A \cap B) = P(A) \times P(B)$
$$= 0.2 \times 0.6$$
$$= 0.12$$

Now 0.4 ≠ 0.12 so the events A and B are not independent. A and B are therefore dependent events.

> Note if the events A and B were independent events, the two probabilities worked out would have been equal.

---

3 In a certain country, 80% of the defendants being tried in the Law Courts actually committed the crime. For those who committed the crime, the probability of being found guilty is 0.9. For those who did not commit the crime, the probability of being found guilty is 0.05.

(a) Find the probability that a randomly chosen defendant is found guilty.

(b) Given that a randomly chosen defendant is found guilty, find the probability that this defendant committed the crime.

Answer

3   (a)   There are two ways in which the defendant could be found guilty.

He/she could have committed the crime and be found guilty.

He/she could have not committed the crime and be found guilty.

Let C = event they committed the crime

G = event they were found guilty

> A tree diagram is drawn. Notice that the first event is whether the person committed the crime or not and the second event is whether they were subsequently found guilty or not.

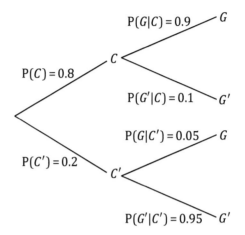

> There are two paths that will find the person guilty. The probability of each path is found and the required probability is the sum of these.

Hence, probability found guilty = $P(C) \times P(G|C) + P(C') \times P(G|C')$

$$= 0.8 \times 0.9 + 0.2 \times 0.05$$

$$= 0.73$$

(b) $P(C \cap G) = P(C)\,P(C|G)$

> Note that $P(C \cap G)$ is the probability that they committed the crime and are found guilty.

$$P(C|G) = \frac{P(C \cap G)}{P(C)} = \frac{0.8 \times 0.9}{0.73} = \frac{72}{73}$$

Step by STEP

A group of university students took part in a survey about student lifestyles. Of these students, 60% claimed to exercise regularly. The probability of a student claiming to exercise regularly and telling the truth is 0.32. The probability of a student claiming to not exercise regularly and telling the truth is 0.97.

(a) Using a tree diagram, or otherwise, show that the probability of a person who does not exercise regularly also claiming they don't exercise regularly is 0.388.

A randomly selected student completes the survey.

(b) Find the probability that the student is telling the truth.

(c) Given that the student is telling the truth, find the probability that the student claims to exercise regularly.

Steps to take

1   You need to understand the question before you start. When you read this question carefully you can see there are two events: claiming to exercise regularly or not and the claim being true or false.

Let the two events be A and B, so the events not happening will be A' and B'.

2 Draw a tree diagram showing all the outcomes and probabilities given in the question.

Use the tree diagram to multiply the probabilities along the path where the student doesn't exercise regularly and is telling the truth that they don't exercise regularly. This should equal 0.388 as given in the question.

3 For part (b) find the branches using the tree diagram where the student is telling the truth. There are two paths to consider:

P(Telling the truth) = P(Exercise regularly) × P(Telling the truth|Exercise regularly)
        + P(Don't exercise regularly) × P(Telling the truth|Don't exercise regularly)

Converting to A and B notation:
$$P(B) = P(A) \times P(B) + P(A') \times P(B)$$

Use the tree diagram to substitute numerical values in for each of these two paths.

4 In part (c), you want to find the probability that the student regularly exercises (A) given that they are telling the truth (B), so this is notated as $P(A|B)$.

Use the conditional probability formula:
$$P(A|B) = \frac{P(A \cap B)}{P(B)}$$

Multiply along the branch of your tree diagram to find the probability that A and B both happen $P(A \cap B)$ and then substitute this value into the formula.

· ·

Answer

(a) Let A = the event that a person claims to exercise regularly

B = the event that the person is telling the truth

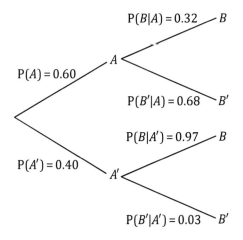

Multiply along for 'Claim don't exercise regularly' and 'Telling the truth given they don't exercise regularly'.

$\big($P(Claim don't exercise regularly) × P(Telling the truth|Claim don't exercise regularly)$\big)$

should equal 0.388.

Converting the events to A and B notation we obtain

$$P(A' \cap B) = P(A') \times P(B|A')$$
$$= 0.40 \times 0.97$$
$$= 0.388$$

(b) $P(B) = P(A) \times P(B|A) + P(A') \times P(B|A')$

$$= 0.60 \times 0.32 + 0.40 \times 0.97$$

$$= 0.58$$

(c) $P(A \cap B) = P(A) \times P(B|A)$

$$= 0.60 \times 0.32$$

$$= 0.192$$

> You found the probability that the student is telling the truth $P(B)$ in part (b), so you can use this in the formula. (i.e. $P(B) = 0.58$)

Now $P(A|B) = \dfrac{P(A \cap B)}{P(B)}$

> Find $P(A|B)$ by substituting your values for $P(A \cap B)$ and $P(B)$ into the conditional probability formula.

$$= \dfrac{0.192}{0.58}$$

$$= 0.33 \text{ (2 s.f.)}$$

1.3 Use of Venn diagrams for conditional probability

One way of solving probability questions is to first draw a Venn diagram. The drawing of Venn diagrams was looked at in Topic 3 of the AS Applied book. You should look back at this topic before continuing with this new topic.

Work through the following example to check your familiarity with Venn diagrams.

Examples

1   The probabilities for two events A and B are shown in the following Venn diagram.

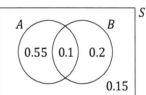

Find:

(a) $P(A)$

(b) $P(A$ only$)$

(c) $P(A \cap B)$

(d) $P(A \cup B)$

(e) $P(A' \cap B)$

Answer

1   (a) P(A) = 0.55 + 0.1 = 0.65

(b) P(A only) = 0.55

(c) P(A ∩ B) = 0.1

(d) P(A ∪ B) = 0.55 + 0.1 + 0.2 = 0.85

(e) P(A' ∩ B) = 0.2

In this section we will look at how we use Venn diagrams to solve problems involving conditional probability.

Suppose we have two events A and B. The probability of B occurring may change depending on whether event A has occurred or not.

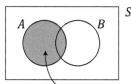

As event A has occured, we only consider the probabilities in set A, i.e. P(A).

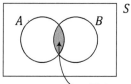

This region represents B occurring given that A has occurred.

Example

1   A and B are two events such that P(A|B) = 0.3, P(A|B') = 0.2 and P(B) = 0.25

Find:

(a) P(A ∩ B)

(b) P(A ∩ B')

(c) P(A)

(d) P(B|A)

(e) P(B|A')

(f) P(B'|A')

Answer

1   (a) $P(A|B) = \dfrac{P(A \cap B)}{P(B)}$

So, P(A ∩ B) = P(A|B) × P(B)

= 0.3 × 0.25

= 0.075

Note that if P(B) = 0.25, then
    P(B′) = 1 − P(B)
    = 1 − 0.25 = 0.75

Once you have enough information you can start to draw the Venn diagram. Always look back at what you have just found to see if it helps to complete the diagram. Here you can see that the answer to part (b) can be added to the Venn diagram.

(b) $P(A \cap B') = P(A|B') \times P(B')$

$\qquad = 0.2 \times 0.75$

$\qquad = 0.15$

(c)

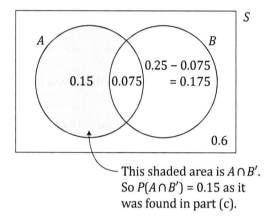

This shaded area is $A \cap B'$.
So $P(A \cap B') = 0.15$ as it was found in part (c).

Using the Venn diagram, P(A) = 0.15 + 0.075

$\qquad = 0.225$

(d) $P(B|A) = \dfrac{P(B \cap A)}{P(A)}$

$\qquad = \dfrac{0.075}{0.225}$

$\qquad = 0.333$

(e) $P(B|A') = 1 − P(B|A)$

$\qquad = 1 − 0.333$

$\qquad = 0.667$

(f) $P(B'|A') = \dfrac{P(B' \cap A')}{P(A')}$

Now $P(A' \cap B') = 1 − (0.15 + 0.075 + 0.175) = 0.6$

Hence $P(B'|A') = \dfrac{P(B' \cap A')}{P(A')}$

$\qquad = \dfrac{0.6}{(0.6 + 0.175)}$

$\qquad = 0.774$

With probability questions, there are often several different methods you can use to solve a problem. Most people find that using a visual method such as using tree or Venn diagrams helps. Sometimes you can use a combination of methods.

2   The events A and B are such that $P(A) = 0.2$, $P(B) = 0.3$, $P(A \cup B) = 0.4$.

(a) Show that A and B are not independent.

(b) Determine the value of:

(i) $P(A'|B)$

(ii) $P(A \cup B')$

Answer using a Venn diagram approach

2   (a)   If events A and B were independent, then $P(A \cap B) = 0.2 \times 0.3 = 0.06$

Now $P(A \cap B) = P(A) + P(B) - P(A \cup B)$

$$= 0.2 + 0.3 - 0.4$$

$$= 0.1$$

A and B are not independent because $0.06 \neq 0.1$

(b)   We can now draw the Venn diagram noting that $P(A \cap B) = 0.1$

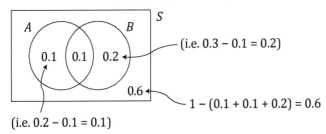

(i.e. $0.3 - 0.1 = 0.2$)

$1 - (0.1 + 0.1 + 0.2) = 0.6$

(i.e. $0.2 - 0.1 = 0.1$)

(i) $P(A'|B) = \dfrac{0.2}{0.1 + 0.2}$

$$= \dfrac{0.2}{0.3}$$

$$= \dfrac{2}{3}$$

(ii) $P(A \cup B')$ is represented by the shaded area shown on the Venn diagram below.

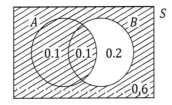

$P(A \cup B') = 0.1 + 0.1 + 0.6$

$$= 0.8$$

$P(A \cup B')$ means everything in A joined onto everything that is not in B.

Answer using an algebraic approach

2   (a)   If events A and B were independent, then $P(A \cap B) = 0.2 \times 0.3 = 0.06$

Now $P(A \cap B) = P(A) + P(B) - P(A \cup B)$

$$= 0.2 + 0.3 - 0.4$$

$$= 0.1$$

A and B are not independent because $0.06 \neq 0.1$

(b)   (i) $P(A \cap B) = P(B)\,P(A|B)$

So $P(A|B) = \dfrac{P(A \cap B)}{P(B)}$

$$= \dfrac{1}{3}$$

This formula can be obtained from the formula booklet.

Now $P(A|B) = 1 - P(A'|B)$

Hence $P(A'|B) = \dfrac{2}{3}$

(ii) $P(A \cup B') = P(A) + P(B') - P(A \cap B')$

$= P(A) + \big(1 - P(B)\big) - \big(P(A) - P(A \cap B)\big)$

$= 0.2 + (1 - 0.3) - (0.2 - 0.1)$

$= \dfrac{4}{5}$ or 0.8

---

3   Events A and B are such that:

$P(A) = 0.2$, $P(B) = 0.4$, $P(A \cup B) = 0.52$.

(a)   Show that A and B are independent.

(b)   Calculate the probability of exactly one of the two events occurring.

(c)   Given that exactly one of the two events occurs, calculate the probability that A occurs.

. . . .

Answer

3   (a)   Using $P(A \cup B) = P(A) + P(B) - P(A \cap B)$

we obtain $P(A \cap B) = P(A) + P(B) - P(A \cup B)$

$= 0.2 + 0.4 - 0.52$

$= 0.08$

> This is the generalised addition law obtained from the formula booklet.

If the events A and B are independent, then probability of both events occurring $= P(A) \times P(B)$

$= 0.2 \times 0.4$

$= 0.08$

As $P(A) \times P(B) = P(A \cap B)$, events A and B are independent.

(b)   The Venn diagram showing P(A only)

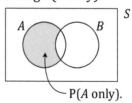

P(A only).

> Note that P(A) would include P($A \cap B$) so P($A \cap B$) has to be subtracted from P(A) to give P(A only).

$P(A \text{ only}) = P(A) - P(A \cap B)$

$= 0.2 - 0.08$

$= 0.12$

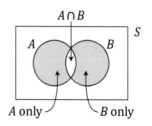

$A \cap B$

A only B only

$$P(B \text{ only}) = P(B) - P(A \cap B)$$

$$= 0.4 - 0.08$$

$$= 0.32$$

$$P(A \text{ or } B \text{ only}) = P(A \text{ only}) + P(B \text{ only}) = 0.12 + 0.32 = 0.44$$

(c) Required probability $= \dfrac{P(A \text{ only occurs})}{P(A \text{ or } B \text{ only occurs})}$

$$= \dfrac{0.12}{0.44}$$

$$= 0.273 \text{ (correct to 3 d.p.)}$$

Produce a list of definitions for the following terms connected with probability:

Active Learning

- Independent events
- Dependent events
- Mutually exclusive events
- Conditional probability.

1.4 Use of two-way tables for conditional probability

Two-way tables are used to work out the number of outcomes so that probabilities can be calculated.

Here is an example of a two-way table showing the numbers of people who like football, tennis, both or neither.

	Tennis	Not tennis	Total
Football	55	15	**70**
Not football	24	26	**50**
Total	**79**	**41**	**120**

If we define event F = liking football and event T = liking tennis we can draw the table using these abbreviations as follows:

	T	T'	Total
F	55	15	**70**
F'	24	26	**50**
Total	**79**	**41**	**120**

Suppose we want to work out the probability of a person chosen at random from the group likes both sports.

We look for the intersection of the row for F and the column for T and then divide this by the total. Hence, we have $P(\text{both sports}) = \dfrac{55}{120} = \dfrac{11}{24}$

If we wanted to find the probability of a person chosen at random liking one sport only, we have $P(\text{liking one sport only}) = \dfrac{24}{120} + \dfrac{15}{120} = \dfrac{39}{120} = \dfrac{13}{40}$

Examples

1

	T	T'	Total
F	55	15	70
F'	24	26	50
Total	79	41	120

Using the above table, work out the probabilities that a person chosen at random from the group:

(a) likes only football,

(b) does not like either sport,

(c) likes football but does not like tennis.

(d) Given that the person chosen likes football, find the probability they do not like tennis.

- -

Answer

1 (a) P(likes only football) $= \dfrac{15}{120} = \dfrac{1}{8}$

(b) P(does not like either sport) $= \dfrac{26}{120} = \dfrac{13}{60}$

(c) P(likes tennis but does not like football) $= \dfrac{24}{120} = \dfrac{1}{5}$

> Remember here that you are choosing out of the people who just like football and not out of the whole group.

(d) P(given they like football, they do not like tennis) $= \dfrac{15}{70} = \dfrac{3}{14}$

In many questions you have to complete the two-way table from information given in the questions. The following example shows this sort of question.

---

2 100 people were asked if they liked cats, dogs, both or neither: 44 said they liked dogs (D), 52 liked cats (C) and of those who liked dogs, 12 also liked cats.

Draw a two-way table to show this information.

- -

Answer

2 In this answer we will work through the processes involved in completing the table.

The number liking both cats and dogs is 12, so add this to the table first.

	D	D'	Total
C	12		
C'			
Total			

We know the total liking cats is 52, so this can be added to the table. We can use this to calculate the number who like cats but don't like dogs.

	D	D'	Total
C	12	40	52
C'			
Total			

The total for liking dogs is 44 so this makes the number liking dogs but not liking cats to be 32. These values are added to the table.

	D	D'	Total
C	12	40	52
C'	32		
Total	44		

The total number of people asked is 100 so this is added to the table.

	D	D'	Total
C	12	40	52
C'	32		
Total	44		100

It is now easy to find the missing values in Total cells.

	D	D'	Total
C	12	40	52
C'	32		48
Total	44	56	100

The missing value can be found, thus completing the table.

	D	D'	Total
C	12	40	52
C'	32	16	48
Total	44	56	100

---

3 A survey was taken by 210 people who were asked if they prefer ordinary or diet cola. The results are shown in this two-way table.

Prefer diet cola	Male	Female	Totals
Yes	65	93	158
No	35	17	52
Totals	100	110	210

If someone surveyed was picked at random, find:

(a) the probability they preferred diet cola

(b) the probability the person is female, given that they preferred diet cola

(c) the probability the person preferred diet cola, given that they are female.

Answer

3 (a) P(Preferred diet cola) $= \dfrac{158}{210} = \dfrac{79}{105}$

(b) It is already known they prefer diet cola so we use the number of respondents who prefer diet cola as the denominator (i.e. 158). The numerator will be the number of females who preferred diet cola (i.e. 93).

P(Female given they preferred diet cola) $= \dfrac{93}{158}$

(c)   We know they are female, therefore we are only picking out of the number of females, so the denominator is 110. The number of females who prefer diet cola is 93.

$$\text{P(Prefer diet cola given they are female)} = \frac{93}{110}$$

1.5 Modelling with probability

A probability model is a mathematical representation of events that are random in nature. It is possible to work out the probabilities of separate events and then use the probability theory we have learned in this topic to find the probabilities of certain scenarios.

Models involving probability are often used in medical research to find whether tests or vaccines are actually working or not. The following example shows how a typical probability model is created, which is then used to answer certain questions. As with all probability questions, you should read the question several times to fully understand the situation being described.

Examples

1   It is known that 4% of a population suffer from a certain disease. When a diagnostic test is applied to a person with the disease, it gives a positive response with probability 0.98. When the test is applied to a person who does not have the disease, it gives a positive response with probability 0.01.

(a)   Using a tree diagram, or otherwise, show that the probability of a person who does not have the disease giving a negative response is 0.9504.

The test is applied to a randomly selected member of the population.

(b)   Find the probability that a positive response is obtained.

(c)   Given that a positive response is obtained, find the probability that the person has the disease.

. . . .

Answer

1   (a)   Let A = the event that a person has the disease.

Let B = the event that a positive response is obtained.

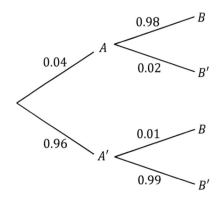

$P(A' \cap B')$ is the probability of not having the disease and giving a negative response.

$$P(A' \cap B') = 0.96 \times 0.99 = 0.9504$$

(b) $P(B) = 0.96 \times 0.01 + 0.04 \times 0.98$

   $= 0.0488$

(c) We need to find the probability of A given that B has occurred (i.e. written as $P(A|B)$)

$$P(A \cap B) = P(B)\,P(A|B)$$

This formula is obtained from the formula booklet.

So $\qquad P(A|B) = \dfrac{P(A \cap B)}{P(B)}$

$$= \dfrac{0.04 \times 0.98}{0.0488}$$

$$= 0.803$$

This is the probability of not having the disease and giving a positive response added to the probability of having the disease and giving a positive response.

---

2 Mary and Jeff are archers and one morning they play the following game. They shoot an arrow at a target alternately, starting with Mary. The winner is the first to hit the target. You may assume that, with each shot, Mary has a probability 0.25 of hitting the target and Jeff has a probability p of hitting the target. Successive shots are independent.

(a) Determine the probability that Jeff wins the game
   (i) with his first shot,
   (ii) with his second shot.

(b) Show that the probability that Jeff wins the game is $\dfrac{3p}{1 + 3p}$

(c) Find the range of values of p for which Mary is more likely to win the game than Jeff.

· ·

Answer

2   (a)   (i)

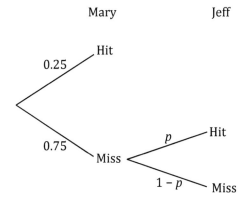

For Jeff to win with his first shot, Mary would have to lose, as she goes first.

P(Jeff wins with 1st shot) = P(Mary misses with 1st shot)
$\qquad\qquad\qquad\qquad\qquad\qquad$ × P(Jeff hits with 1st shot)

$\qquad\qquad\qquad = 0.75 \times p = 0.75p$

(ii)

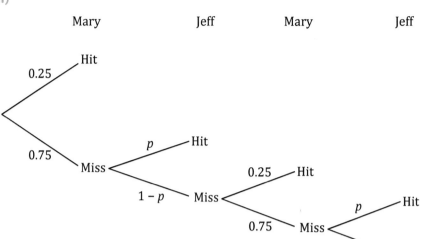

$$P(\text{Jeff wins with 2nd shot}) = \begin{array}{l} P(\text{Mary misses with 1st shot}) \times \\ P(\text{Jeff misses with 1st shot}) \times \\ P(\text{Mary misses with 2nd shot}) \times \\ P(\text{Jeff hits with 2nd shot}) \end{array}$$

Prob Jeff wins with 2nd shot $= 0.75 \times (1 - p) \times 0.75 \times p$
$$= 0.75^2(1 - p)p$$

(b) Prob Jeff wins with 3rd shot $= 0.75 \times (1 - p) \times 0.75 \times (1 - p) \times 0.75 \times p$
$$= 0.75^3(1 - p)^2p$$

A geometric series is forming as follows:

$$0.75p + 0.75^2(1 - p)p + 0.75^3(1 - p)^2p + \dots$$

Note that this is an infinite geometric series.

Common ratio, $r = 0.75(1 - p)$

$$S_\infty = \frac{a}{1 - r} \text{ provided that } |r| < 1$$

Now a is the first term of the sequence, so $a = 0.75p$

Hence $S_\infty = \dfrac{a}{1 - r} = \dfrac{0.75p}{1 - 0.75(1 - p)} = \dfrac{0.75p}{1 - 0.75 + 0.75p} = \dfrac{0.75p}{0.25 + 0.75p}$

Dividing top and bottom by 0.25, we obtain

Probability that Jeff wins the game $= \dfrac{3p}{1 + 3p}$

(c) Mary is more likely to win if this probability is less than 0.5 (because Mary's probability of winning would be greater than, or equal to, 0.5).

Hence, $\dfrac{3p}{1 + 3p} < 0.5$

$$3p < 0.5 + 1.5p$$
$$1.5p < 0.5$$
$$p < \frac{1}{3}$$

Test yourself

1 Using set notation, describe the following shaded areas in each of these Venn diagrams.

(a)
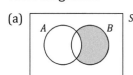

(b)
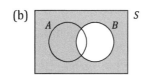

(c)
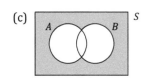

(d)
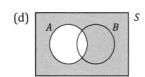

[4]

2 The two-way table below shows the preferences for a meal out for 50 students.

	Indian	Chinese	Total
Male	17	14	**31**
Female	12	7	**19**
Total	**29**	**21**	**50**

Using the two-way table, work out for a person chosen at random from the 50 students.
(a) The probability that they prefer Indian food. [1]
(b) The probability they are female and prefer Chinese food. [2]
(c) The probability they prefer Indian food, given that they are male. [2]

3 A card is chosen at random from a pack of 52 playing cards. If R is the event that a card is a red card and P is the event that a card is a picture card.
(a) Complete the following two-way table to show this information.

	R	R'	Totals
P			
P'			
Totals			

[2]

(b) A card is chosen at random from the pack. Find:
   (i) $P(P \cap R)$ [1]
   (ii) $P(P' \cap R')$ [1]
   (iii) $P(P|R)$ [1]
   (iv) $P(P'|R)$. [1]

4 The Venn diagram shown shows the probabilities for events A and B.

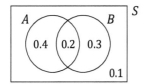

(a) Show that events A and B are not independent. [2]

(b) Find:
    (i) $P(A')$ [1]
    (ii) $P(A \cup B)'$ [1]
    (iii) $P(A' \cap B)$ [1]
    (iv) $P(A|B)$ [1]
    (v) $P(A'|B)$.   [1]

5 The events A and B are such that:
    $P(A) = 0.3$, $P(B) = 0.4$.
Evaluate $P(A \cup B)$ in each of the following cases.
(a) A and B are mutually exclusive.   [2]
(b) A and B are independent.   [3]
(c) $P(A|B) = 0.25$.   [4]

6 The events A and B are such that:
    $P(A) = 0.4$, $P(B) = 0.5$ and $P(A \cup B) = 2 \times P(A \cap B)$.
(a) Show that $P(A \cap B) = 0.3$.   [2]
(b) Evaluate $P(A|B)$.   [2]
(c) Evaluate $P(B|A')$.   [3]

7 There are 120 students in a sixth form taking science subjects: 70 students take chemistry (C) and 61 students take physics (P). Of those who take chemistry, 21 also take physics.
(a) Draw a two-way table to show the information above.   [3]
(b) Draw a Venn diagram to illustrate the information shown in the two-way table.   [3]
(c) One student is selected at random from the group taking science subjects. Find:
    (i) $P(C')$ [1]
    (ii) $P(C' \cap P')$ [1]
    (iii) $P(P|C)$ [1]
    (iv) $P(C'|P)$.   [1]

8 The events A and B are such that $P(A) = 0.3$, $P(B) = 0.5$ and $P(A \cup B) = 0.7$
(a) Show that the events A and B are not independent.   [3]
(b) Determine the value of:
    (i) $P(A'|B)$ [2]
    (ii) $P(A' \cup B)$.   [2]

Summary

Check you know the following facts:

The multiplication law for independent events

If events A and B are independent:

$$P(A \cap B) = P(A) \times P(B)$$

The As and Bs can be swapped around in this formula and you can also swap A for A' and B for B' for all combinations.

The multiplication law for dependent events

If events A and B are dependent:

$$P(A \cap B) = P(A) \times P(B|A)$$

The generalised addition law

$$P(A \cup B) = P(A) + P(B) - P(A \cap B)$$

The generalised addition law can be used for dependent or independent events.

This formula is included in the formula booklet.

The conditional probability formula

$P(A|B)$ means the probability of A given that B has occurred.

$$P(A|B) = \frac{P(A \cap B)}{P(B)}$$

This formula is included in the formula booklet.

The As and Bs can be swapped around in this formula and you can also swap A for A' and B for B' for all combinations.

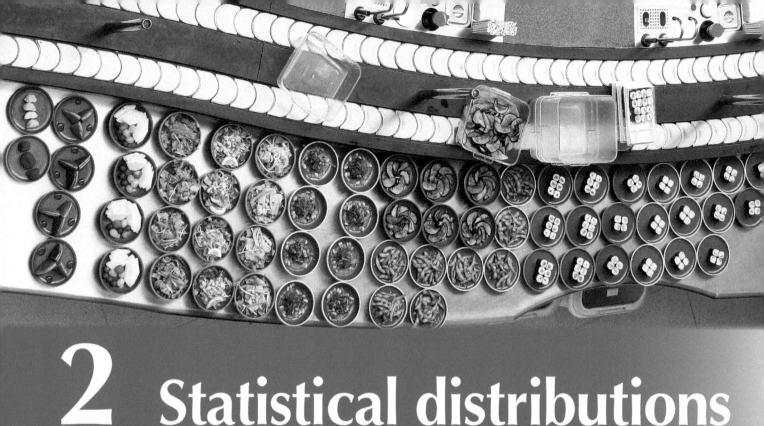

2 Statistical distributions

Introduction

There are several statistical distributions that can be used to work out probabilities and you have to first consider whether the data being looked at are continuous or discrete.

Discrete data jump from one value to the next without in-between values (e.g. the scores obtained when throwing six-sided dice, the number of children in a family), whereas continuous data have no gaps between data values (e.g. the length of pieces of string, the heights of students in a college, etc.).

There are three discrete probability distributions: discrete uniform, binomial and Poisson. These three distributions were covered in the AS Applied book, so you will need look back at Topic 4 on pages 69 to 86 to refresh your memory.

There are two continuous probability distributions covered in this topic and they are: continuous uniform and normal. Continuous distributions were not covered in the AS Applied Unit, so all this material will be new.

The continuous uniform distribution is a probability distribution where all values in the same interval of allowed values, have equal probabilities of occurring.

The normal distribution is a continuous distribution meaning all values are possible, so there are no gaps between values. The normal distribution is very useful as it enables you to find the probability of a quantity taking certain values.

This topic covers the following:

2.1 Continuous uniform distributions

2.2 The 'bell-shaped curve' of the normal distribution

2.3 The standard normal distribution

2.4 Using the normal distribution tables or a calculator to work out probabilities

2.5 Selecting an appropriate probability distribution for a particular context

2.1 Continuous uniform distributions

The discrete uniform distribution was covered on page 84 of the AS Applied book and it applied to discrete data. Here we will be looking at the continuous uniform distribution that applies to continuous data.

The continuous uniform distribution is often called the rectangular distribution because of the shape of its graph. Any intervals of equal length are equally likely to occur between a minimum value a and a maximum value b. The probability of values less than a or greater than b is zero.

The distribution is often abbreviated as U[a, b].

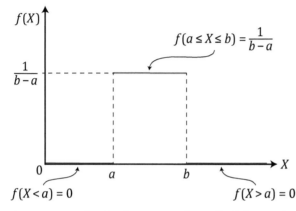

The graph above shows a graph of a function of X called the probability density function, which is a measure of the probability of the random variable X falling within a particular range of values.

If $X \sim$ U[a, b], then:

Mean, $E(X) = \frac{1}{2}(a + b)$

Variance, $Var(X) = \frac{1}{12}(b - a)^2$

Both of these formulae are included in the formula booklet.

Using the continuous uniform distribution to work out probabilities

The continuous uniform distribution can be used to work out probabilities in the following way.

Suppose we want to find the probability of a value of X being between two values, d and c, i.e. $P(c \leq X \leq d)$. We can show these values along with the values of a and b on the following graph:

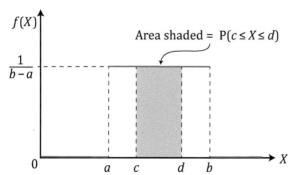

The following formula can be used to work out the shaded area which is $P(c \le X \le d)$.

$$P(c \le X \le d) = \frac{d - c}{b - a}$$

Examples

1   The time in minutes to be served at a busy bar follows a continuous uniform distribution over the interval $[3, 9]$.

Find:

(a)   The expected wait in minutes at the bar waiting to be served.

(b)   The variance in minutes at the bar waiting to be served.

(c)   The probability that a customer waits more than 7 minutes to be served.

. . . .

Answer

1   (a)   Mean, $E(X) = \frac{1}{2}(a + b) = \frac{1}{2}(3 + 9) = 6$ minutes

(b) $Var(X) = \frac{1}{12}(b - a)^2 = \frac{1}{12}(9 - 3)^2 = 3$ minutes

(c)
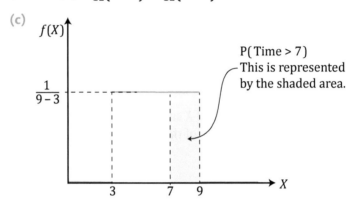

$$P(\text{Time} > 7) = \frac{d - c}{b - a} = \frac{9 - 7}{9 - 3} = 0.33 \ (2 \text{ d.p.})$$

---

2   The waiting time for a train that leaves every 20 minutes can be modelled as a continuous uniform distribution from 0 to 20 minutes. Find the probability that a person arriving at a random time will wait between 5 and 10 minutes.

. . . .

Answer

2    $P(5 < x < 10) = \left(\frac{(10 - 5)}{(20 - 0)} \right)$

$= 0.25$

---

3   A piece of string of length 20 cm is cut at a random point. The length of the longer piece is denoted by X cm and the length of the shorter piece is denoted by Y cm. You may assume that X is uniformly distributed on the interval $[10, 20]$.

(a)   Determine $P(Y < 8)$.

(b)   (i)   Express Y in terms of X.

(ii)   Determine $P(XY > 64)$.

Answer

3   (a) $P(Y < 8) = P(X > 12)$

$$= \frac{d - c}{b - a}$$

$$= \frac{20 - 12}{20 - 10}$$

$$= 0.8$$

(b)   (i) $Y = 20 - X$

(ii) $P(XY > 64) = P[X(20 - X) > 64]$

$$= P(X^2 - 20X + 64 < 0)$$

$$= P\big((X - 4)(X - 16) < 0\big)$$

Now $(X - 4)(X - 16) < 0$

The critical values are 4 and 16

> This is a quadratic inequality. The graph of the function will be U-shaped and the curve intersects the x-axis at 4 and 16.

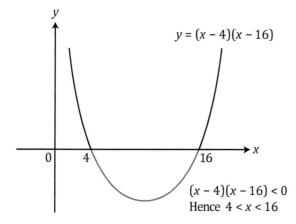

$y = (x - 4)(x - 16)$

$(x - 4)(x - 16) < 0$
Hence $4 < x < 16$

X cannot be ≤ 10 as it wouldn't be the longer piece.

Hence $10 < X < 16$.

The required region is $P(10 < X < 16) = \dfrac{d - c}{b - a} = \dfrac{16 - 10}{20 - 10} = 0.6$

Hence $P(XY > 64) = 0.6$

2.2 The 'bell-shaped curve' of the normal distribution

If you investigated the heights of a large number of 18-year-olds and found that the mean height was 1.75 m, you would expect as many of them to be below this height as above. You would conclude that the distribution was symmetrical either side of the mean. If you plotted all the heights against the likelihood of them, the graph would be bell-shaped and symmetrical about the mean.

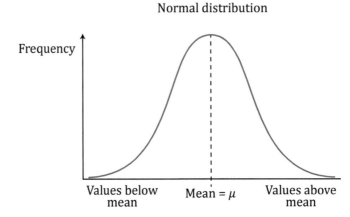

Notice that the maximum of this curve is the mean, μ, called the population mean as it is based on the entire group rather than just a smaller sample. All normal distributions are bell-shaped but they can be steeper or less steep depending on the variance (σ^2). The curve has asymptotes at each end and is symmetrical at the mean, median and mode which all lie at the same point.

The curve has points of inflexion at $\mu + \sigma$ and $\mu - \sigma$. At $\mu - \sigma$, which is to the left of the mean, the curve changes from convex (i.e. above its own tangent) to concave (i.e. below its own tangent). At $\mu + \sigma$, on the other side of the mean, the curve changes from concave to convex.

Here are some facts about normal distributions:

- They are used to model the probability of a continuous random variable.
- They are continuous probability distributions.
- The total area under the curve is always equal to 1.

The normal distribution uses two parameters, the population mean μ and the population variance σ^2.

> The approximately sign ~ means 'is distributed as'.

If X is a normally distributed variable then it can be written as $X \sim N(\mu, \sigma^2)$

Suppose we want to find the probability $P(X < x)$, we can add up the probability of all the values less than x. This is represented by the area under the curve to the left of x.

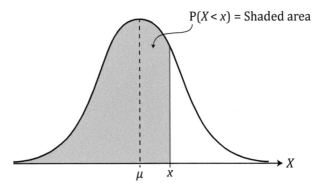

There are many different normal distribution curves each with a different mean and variance, so we instead use a standardised normal distribution and adjust our parameters to fit the curve.

2.3 The standard normal distribution

The standard normal distribution is centred at 0 and has a variance and hence standard deviation of 1.

We can write this as $Z \sim N(0, 1)$

We now need to adjust the parameters so they fit this curve.

We change the variable from X to Z and instead of x we now use z. Values can be changed from X values to Z values using the formula:

$$z = \frac{x - \mu}{\sigma}$$

> As $Z \sim N(0, 1)$, $\mu = 0$ and $\sigma^2 = 1$. σ^2 is the variance so σ is the standard deviation.

The z-values change the normal distribution with mean μ and standard deviation σ to the standard normal distribution which has a mean of 0 and standard deviation of 1.

Once an x-value has been converted to a z-value using the formula, we can draw the graph like this:

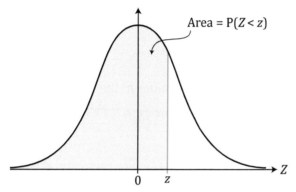

> Remember that the standard normal distribution is always centred on 0.

2.4 Using the normal distribution tables or a calculator to work out probabilities

In this section you will be looking at how tables or a calculator can be used to work out probabilities using either the normal distribution or the standard normal distribution.

Example

1 If $Z \sim N(0, 1)$ find $P(Z > 1.5)$

Now $N(0, 1)$ is a standard normal distribution, so we do a quick sketch showing the Z-value and the shaded area representing the probability. The first graph shows this.

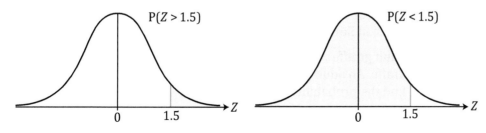

Now the tables show the probability less than a certain Z-value which is shown in the second diagram. Now as the total area under the standard normal distribution is 1, we can use

$$P(Z > 1.5) = 1 - P(Z < 1.5)$$

Now $P(Z < 1.5)$ can be found from tables.

From tables $P(Z < 1.5) = 0.93319$

Hence, $P(Z > 1.5) = 1 - P(Z < 1.5) = 1 - 0.93319 = 0.06681$

Important note

When dealing with a continuous distribution you can use $>$ or $\geq$ interchangeably and you can also use $<$ or $\leq$ interchangeably so don't worry about this. This is due to the fact that the probability of a particular value is zero so, for example, $P(X < 5)$ is the same as $P(X \leq 5)$ as $P(X = 5) = 0$. This makes sense as there is no corresponding area under the curve for a single value such as $X = 5$ and it is the area that is equal to the probability.

Examples

1   If $Z \sim N(0, 1)$ find $P(Z < -2)$

Note that there are no negative Z-values in the tables.

. . . .

Answer

1

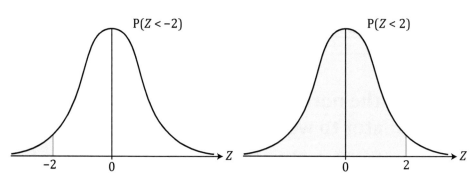

Here we use the symmetry of the graphs, so

$$P(Z < -2) = P(Z > 2) = 1 - P(Z < 2)$$

Using tables $P(Z < -2) = 1 - 0.97725 = 0.02275$

---

2   The heights of adult giraffes are normally distributed with mean 550 cm and standard deviation 20 cm.

    (a)   Find the probability that a randomly selected adult giraffe has a height greater than 560 cm.

    (b)   Any adult giraffe whose height is above 560 cm is classified as being a tall giraffe. An adult giraffe is selected at random. Given that this giraffe is tall, find the probability that its height is greater than 580 cm.

Answer

2   (a)   Let X be the random variable 'height of an adult giraffe'. X is normally distributed with mean 550 cm and standard deviation 20 cm, so:

$$X \sim N(550, 20^2)$$

Sketch graphs showing the distribution for the random variable, X, and the standardised variable, Z:

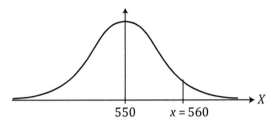

We need to find $P(X > 560)$.

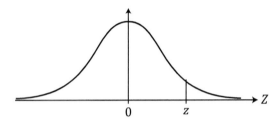

The following formula is used to find the standardised variable z:

$$z = \frac{x - \mu}{\sigma}$$

$$z = \frac{560 - 550}{20}$$

$$= 0.5$$

So $P(X > 560) = P(Z > 0.5)$

Now $P(Z > 0.5) = 1 - P(Z < 0.5)$ because the area under the standardised variable graph is equal to 1 and the normal distribution tables only give areas to the left of the line and you want to find the area to the right.

So using the normal distribution table in the statistical table booklet, you should find that:

$$P(Z < 0.5) = 0.69146$$

So $1 - P(Z < 0.5) = 1 - 0.69146$

$$= 0.309 \text{ (3 s.f.)}$$

(b)   You want to find the conditional probability that the giraffe is taller than 580 cm, given that is taller than 560 cm, so we would notate this as:

$$P(X > 580 | X > 560)$$

Using the conditional probability formula we obtain:

$$P(A|B) = \frac{P(A \cap B)}{P(B)}$$

so $P(X > 580 | X > 560) = \dfrac{P(X > 580 \cap X > 560)}{P(X > 560)}$

If the giraffe is taller than 580 cm, it will be taller than 560 cm so therefore $P(X > 580 \cap X > 560) = P(X > 580)$

Hence we have $P(X > 580|X > 560) = \dfrac{P(X > 580)}{P(X > 560)}$

Converting to the standardised variable, Z, using the formula:

$$z = \frac{x - \mu}{\sigma}$$

Substituting in the values for the random variable, X, the mean, μ and the standard deviation, σ, to find the z value for $P(X > 580)$:

$$z = \frac{580 - 550}{20}$$

$$= 1.5$$

Rewriting the formula in terms of z, and also substituting in the value for $P(X > 560)$ found in part (a), we obtain:

$$P(X > 580|X > 560) = \frac{P(Z > 1.5)}{0.309}$$

The normal distribution table only gives areas to the left of the line and we want to find the area to the right, so like in part (a), we subtract $P(Z < 1.5)$ from 1 to find $P(Z > 1.5)$:

$$P(X > 580|X > 560) = \frac{1 - P(Z < 1.5)}{0.309}$$

Using the normal distribution table in the statistical tables booklet we obtain:

$$P(Z < 1.5) = 0.93319$$

so $1 - P(Z < 1.5) = 1 - 0.93319 = 0.06681$

Substituting this value back into the original formula, we obtain:

$$P(X > 580|X > 560) = \frac{0.06681}{0.309} = 0.216 \text{ (3 s.f.)}$$

Use of calculators instead of tables

The WJEC Specification states that calculators must have certain functionalities including the ability to compute summary statistics and access probabilities from standard statistical distributions. You must learn how to use the calculator because you may get questions which cannot easily be answered using the tables.

Active Learning

Work through the text and learn how you would work out the values using your calculator. If you get stuck then have a look at some of the YouTube videos on using your model of calculator. Remember that the answers you obtain using a calculator may be slightly different than those worked out using tables.

Here are a few examples of using a calculator to give you some practice.

Example

1   The marks for a maths test for a class of students are normally distributed with a mean of 75 marks and a standard deviation of 10 marks. What is the probability of a randomly picked student obtaining a test mark above 60?

Answer

1   We can draw the following graph:

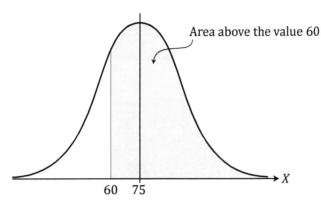

The following formula is used to find the standardised variable z:

Here $x = 60$, $\mu = 75$ and $\sigma = 10$.

$$z = \frac{x - \mu}{\sigma}$$

$$z = \frac{60 - 75}{10}$$

$$= -1.5$$

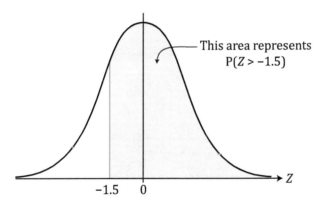

This area is the same as shown below

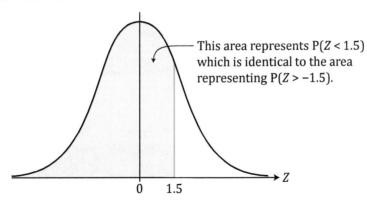

Looking up this z-value (i.e. 1.5) in the Normal distribution table, we obtain:

Probability of a mark greater than 60 = 0.93319

Note that a mark of above 60 corresponds to a z-value of −1.5.

Using a calculator to solve this example

We are asked to find $P(X > 60)$.

1   Select 'Distribution' – you will see a bell-shaped curve like the one selected in the display shown below.

2   You now choose a distribution. The distribution we need is 'Normal CD'.

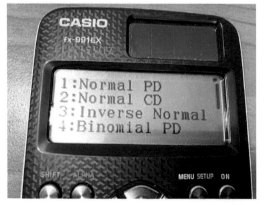

Be careful – make sure you use the Normal CD which stands for Normal Cumulative distribution function.

3   You now have to enter the parameters, but before you do it is always advisable to draw a quick sketch of the distribution and shade the area that represents the required probability.

Hence we produce the following sketch:

We now enter the following parameters into the calculator:

- Lower = 60 (this is the lowest value for the shaded area).
- Upper = 1×10^{99} (remember that this graph on the upper-tail approaches infinity so we just put a very large number in as the upper value).
- $\sigma = 10$ (this is the standard deviation of the distribution).
- $\mu = 75$ (this is the mean value for the distribution).

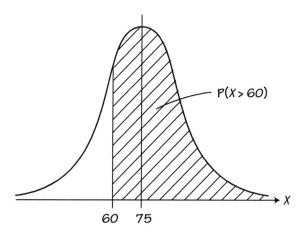

4   The answer is now displayed

Probability of a mark greater than 60 = 0.93319

Finding the probability that a normally distributed random variable lies below a certain value

If $X \sim N(50, 4^2)$ then find, using a calculator, $P(X < 55)$.

First draw a quick sketch of the distribution:

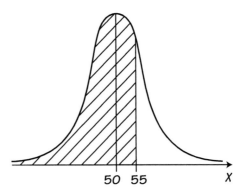

Now set your calculator to 'Normal CD' and enter the following parameters into the calculator:

- Lower = 1×10^{-99} (remember that this graph on the lower-tail approaches a very small value for X so we just put a very small number in as the lower value).

- Upper = 55 (this is the highest value).

- $\sigma = 4$ (this is the standard deviation of the distribution).

- $\mu = 50$ (this is the mean value for the distribution).

Hence $P(X < 55) = 0.89435$

Finding the probability that a normally distributed random variable lies between a range of values

If $X \sim N(150, 20^2)$ then find, using a calculator, $P(120 < X < 170)$.

First draw a quick sketch of the distribution:

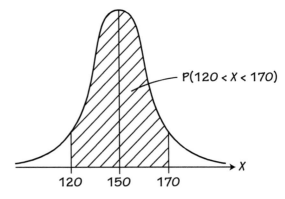

Now set your calculator to 'Normal CD' and enter the following parameters into the calculator:

- Lower = 120 (this is the lowest value of the range).

- Upper = 170 (this is the highest value of the range).

- $\sigma = 20$ (this is the standard deviation of the distribution).
- $\mu = 150$ (this is the mean value for the distribution).

Hence $P(120 < X < 170) = 0.77454$

Finding the probability that a normally distributed random variable lies below a certain value or above a certain value

If $X \sim N(15, 2^2)$ then find, using a calculator, $P(X < 10 \text{ or } X > 16)$, giving your answer to three decimal places.

First draw a quick sketch of the distribution:

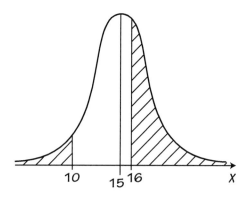

We need to find the two areas representing the probabilities and then finally add them together to give the required probability (i.e. $P(X < 10 \text{ or } X > 16)$)

Considering the lower tail to find $P(X < 10)$, we have the following parameters which we enter into the calculator in its 'Normal CD' setting:

- Lower = 1×10^{-99} (remember that this graph on the lower tail approaches a very small value for X so we just put a very small number in as the lower value).
- Upper = 10 (this is the highest value).
- $\sigma = 2$ (this is the standard deviation of the distribution).
- $\mu = 15$ (this is the mean value for the distribution).

Hence $P(X < 10) = 0.006210$

Considering the upper tail to find $P(X > 16)$, we have the following parameters which we enter into the calculator in its 'Normal CD' setting:

- Lower = 16.
- Upper = 1×10^{99} (this has to be a very large value).
- $\sigma = 2$ (this is the standard deviation of the distribution).
- $\mu = 15$ (this is the mean value for the distribution).

Hence $P(X > 16) = 0.308538$

$P(X < 10 \text{ or } X > 16) = 0.006210 + 0.308538$

$= 0.314748$

$= 0.315$ (3 d.p.)

This is the OR law for probability. The two probabilities are added together.

Using the Normal distribution tables backwards

Up to now we have used the X-value along with the mean μ and standard deviation σ to work out the z-value and then used the tables to work out the probability.

In some questions you will be given the probability and then have to use the table backwards to work out the z-value. Once the z-value has been found it can then be used to find a missing value, such as the standard deviation, mean or x-value.

This is best understood by following these examples.

Examples

1 X is a normally distributed, random variable with a mean of 20 and a standard deviation of 3.

Find the value of x, correct to two decimal places such that:

(a) $P(X < x) = 0.75$

(b) $P(X > x) = 0.3$

(c) $P(18 < X < x) = 0.4$

> It is important to note that if any of these inequalities had contained an equals sign (i.e. $\leq$ or $\geq$) then the working out would have been identical.

Answer

1 (a) We first find the z-value by looking for a probability of 0.75 in the main body of the table for the normal distribution function and then reading off the corresponding z-value.

Look for a value for the probability as near as possible to 0.75. Don't worry if it is not exactly 0.75.

From the table we find $P(X < x) = 0.75$ corresponds to a z-value of 0.67.

We can now use the formula to work out the value of x that corresponds to this z-value.

$$z = \frac{x - \mu}{\sigma}$$

$$0.67 = \frac{x - 20}{3}$$

$$x = 22.01$$

> Note that when you look for 0.75 in the table you have the two values 0.74857 and 0.75175. As 0.74857 is nearer to 0.75 we choose this value and read off its z-value, which is 0.67.

(b) $P(X > x) = 0.3$

It is a good idea to produce a sketch of the distribution of the random variable X to show where the area representing $P(X > x) = 0.3$ lies.

> Note that $P(X > x) = 0.3$ must be in the upper tail, but the tables give probabilities for $P(X < x)$, so we need to subtract 0.3 from 1.

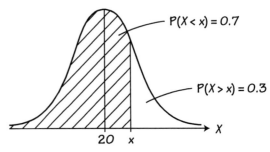

The area representing $P(X > x) = 0.3$ will be at the upper tail.

As the tables are for $P(X < x)$ we can change this area/probability to $1 - 0.3 = 0.7$

Now we use the z-values

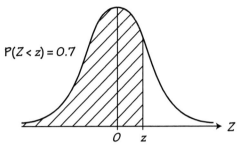

$$P(Z < z) = 0.7$$

A probability of 0.7 is looked up in the body of the Normal distribution table and the corresponding z-value was found to be 0.52 .

$$z = \frac{x - \mu}{\sigma}$$

$$0.52 = \frac{x - 20}{3}$$

$$x = 21.56$$

> We are using this formula to find the value of x.

(c) It is best to draw a sketch of the random variable X first to work out where the area corresponding to a probability of 0.4 lies.

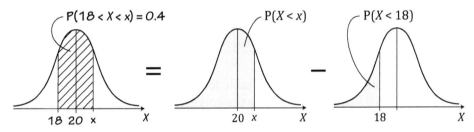

$$P(18 < X < x) = 0.4 \qquad P(X < x) \qquad P(X < 18)$$

Using the diagram you can see that $P(18 < X < x) = P(X < x) - P(X < 18)$

Now we need to find the z-value for $X = 18$

$$z = \frac{x - \mu}{\sigma}$$

$$z = \frac{18 - 20}{3} = -0.67$$

Hence $P(z < -0.67) = 1 - P(z < 0.67) = 1 - 0.74857 = 0.25143$

$$0.4 = P(Z < z) - 0.25143$$

$$P(Z < z) = 0.65143$$

Now we use the table backwards to look up the probability and find the corresponding z-value.

The formula $z = \frac{x - \mu}{\sigma}$ can then be used to find the value of x.

From the table, the nearest value to 0.65143 is 0.65173 and this gives a z-value of 0.39.

Using $z = \frac{x - \mu}{\sigma}$

we have $0.39 = \frac{x - 20}{3}$

Hence $x = 21.17$

BOOST
Grade ⇧⇧⇧⇧

Always look back at the question to see if any accuracy is required for your answer. Here the value is to be given to two decimal places.

2   The marks for 100 maths students in a GCSE exam are normally distributed with a mean mark of 45 and a standard deviation of 20 marks. The mark, x, to obtain a level 9 is to be set so that 10% of students obtain a level 9.

Estimate the value of the mark x giving your answer to the nearest integer.

Answer

2   From the percentage, we know the probability of a randomly picked student obtaining a level 9 is 0.1 .

Hence $P(X > x) = 0.1$

We need the mark such that $P(X < x) = 0.9$

The standard normal curve can be drawn as follows:

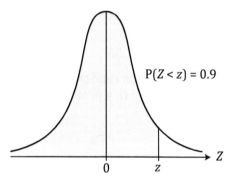

We now look in the body of the table for a probability value as near as possible to 0.9 and then look for the corresponding z-value.

The nearest value in the table to 0.9 is 0.89973 which gives a z-value of 1.28.

We now substitute this z-value into $z = \dfrac{x - \mu}{\sigma}$ to find the x-value.

$$z = \frac{x - \mu}{\sigma}$$

$$1.28 = \frac{x - 45}{20}$$

$$x = 70.6$$

Minimum number of marks to obtain a level 9 = 71 marks.

Using the normal distribution backwards using a calculator

You can use a calculator to find the value of x corresponding to a certain probability. You have to set your calculator to 'Distributions' and then 'Inverse Normal'.

You then enter the following parameters:

● Area (this is the same as the probability).

● The standard deviation, σ.

● The mean, μ.

The following examples show this method.

Examples

1   If $X \sim N(30, 36)$ and $P(X < x) = 0.95$ find x using a calculator, giving your answer to two decimal places.

> N(30, 36) means the mean is 30 and the variance is 36, so the standard deviation is 6.

Answer

1   Note that as $P(X < x) = 0.95$ and the areas are worked out from the left tail and 0.95 is past 0.5, it means that the value of x is to the right of the mean.

We can sketch the distribution:

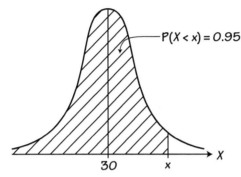

Setting the calculator to 'Inverse Normal' and then entering the following parameters:

- Area = 0.95 (this is the same value as the probability)
- $\sigma = 6$
- $\mu = 30$

Using the calculator, $x = 39.8691$

$x = 39.87$ (2 d.p.)

---

2   It was found that the time taken, T min, to fit a new tyre on a car, can be modelled as a normally distributed variable with a mean of 23 minutes and a standard deviation of 6.5 minutes.

Using **a calculator**, find:

(a) $P(T > 30)$

(b) $P(15 \leq T \leq 25)$

(c)   The value of T, to two decimal places, such that $P(T \leq t) = 0.25$.

Answer

2   (a)   We now enter the following parameters into the calculator:

- Lower = 30 (this is the lowest value for the shaded area).
- Upper = 1×10^{99} (remember that this graph on its upper tail approaches infinity so we just use a very large number as the upper value).
- $\sigma = 6.5$ (this is the standard deviation of the distribution).
- $\mu = 23$ (this is the mean value for the distribution).

Hence $P(T > 30) = 0.14076$

(b) We now enter the following parameters into the calculator:

- Lower = 15
- Upper = 25
- $\sigma = 6.5$ (this is the standard deviation of the distribution)
- $\mu = 23$ (this is the mean value for the distribution)

Hence $P(15 \leq T \leq 25) = 0.51164$

(c) Remember to change the calculator to 'Inverse Normal' as you are working backwards from a given probability to a certain value of T in this case.

You are asked for the area which is the same as the probability, so this is 0.25 (note that as this is less than 0.5 this probability will be in the lower tail).

We can now sketch the distribution like this:

It is always worth drawing a sketch so you can see which tail you are investigating.

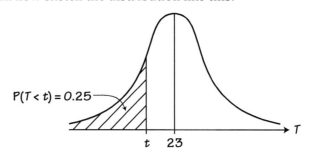

We now enter the following parameters into the calculator:

- Area = 0.25
- $\sigma = 6.5$
- $\mu = 23$

Hence $t = 18.6158$ minutes

Using a calculator to calculate z-values

A calculator can be used to work out z-values: z-values refer to the standard normal distribution where $Z \sim N(0, 1)$.

Example

1 If $X \sim N(50, 25)$, find $P(45 < X < 55)$ by first converting to the standard normal distribution and finding the z-values.

. . . .

Answer

1 $Z \sim N(0, 1)$ as the standard normal distribution is being used.

Converting the x-values to z-values, we have:

When $x = 45$, $\quad z = \dfrac{x - \mu}{\sigma} = \dfrac{45 - 50}{5} = -1$

When $x = 55$, $\quad z = \dfrac{x - \mu}{\sigma} = \dfrac{55 - 50}{5} = 1$

Hence we need to find $P(-1 < Z < 1)$ and here we use the following parameters for the standard normal distribution.

- Lower value = −1
- Upper value = 1

- $\sigma = 1$ (this is the standard deviation for all standard normal distributions)
- $\mu = 0$ (this is the mean value for all standard normal distributions)

Hence $P(45 < X < 55) = 0.6827$

Finding the mean, μ of a normal distribution using a calculator

To find the mean μ of a normal distribution using a calculator you have to first find the z-value and then use the formula $z = \dfrac{x - \mu}{\sigma}$ to find the value of μ.

Example

1   The time taken, X minutes, to perform an MOT at a garage can be modelled as a random variable using a normal distribution with mean μ minutes and standard deviation 10 minutes.

If $P(X < 55) = 0.8$, find the value of μ correct to two decimal places.

· ·

Answer

1   We need to first use the probability of 0.8 to find the z-value. As the z-value refers to the standard normal distribution we use $Z \sim N(0, 1)$.

We can draw the following sketch of the standard normal distribution:

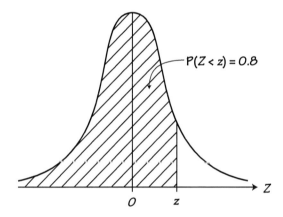

$P(Z < z) = 0.8$

Remember that if a question involves the use of z-values you must use a mean of 0 and a standard deviation of 1 as you will need to use the standard normal distribution.

Use the calculator and select 'Inverse Normal' and then enter the following parameters:

- Area = 0.8
- $\sigma = 1$
- $\mu = 0$

The calculator gives a z-value of 0.8416

We now use the formula with $x = 55$, $\sigma = 10$ to find the value of the mean, μ.

$$z = \frac{x - \mu}{\sigma}$$

$$0.8416 = \frac{55 - \mu}{10}$$

Mean, $\mu = 46.58$

Finding the standard deviation, σ of a normal distribution using a calculator

To find the standard deviation of a normal distribution using a calculator you have to first find the z-value and then use the formula $z = \dfrac{x - \mu}{\sigma}$ to find the value of σ.

Example

1 If $X \sim N(30, \sigma^2)$ and $P(X < 40) = 0.9$ find σ using a calculator, giving your answer to two decimal places.

Answer

1 First we use the probability of 0.9 to find the z-value. As the z-value refers to the standard normal distribution, we use $Z \sim N(0, 1)$.

We can draw the following sketch of the standard normal distribution:

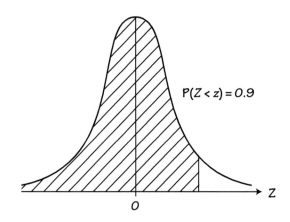

Use the calculator and select 'Inverse Normal' and then enter the following parameters:

- Area = 0.9
- $\sigma = 1$
- $\mu = 0$

The calculator gives a z-value of 1.2816

We now use the formula with $x = 40$, $\mu = 30$ to find the value of the standard deviation, σ.

$$z = \frac{x - \mu}{\sigma}$$

$$1.2816 = \frac{40 - 30}{\sigma}$$

Standard deviation, $\sigma = 7.80$

Finding the mean, μ, and standard deviation, σ, of a normal distribution using a calculator

The approach for finding the mean, μ, and standard deviation, σ, is similar to finding their values separately except this time as there are two unknowns so it is necessary to find two equations and then solve these simultaneously.

Don't get this standard deviation mixed up with the standard deviation which refers to the standard deviation of X. Here we are using the standard normal distribution which always has a standard deviation of 1.

Step by STEP

The random variable X can be modelled using a normal distribution having mean μ and standard deviation σ.

If $P(X < 12) = 0.3$ and $P(X > 15) = 0.12$, find the value of μ and σ, giving your answers correct to 3 decimal places.

Steps to take

1   First decide whether you are going to use tables or a calculator to solve this. Either method is good but here we will use a calculator.

2   Sketch the normal distribution for X and shade the areas representing the two probabilities.

3   Convert this normal distribution to the standard normal distribution (i.e. $Z \sim N(0, 1)$).

    Use the 'Inverse Normal' on the calculator and find the z-value for each of the tails of the distribution.

4   Use the two probabilities and two z-values along with the formula to produce two equations.

5   Solve the two equations simultaneously to find the values of μ and σ.

6   Give the answers correct to 3 decimal places.

. . . .

Answer

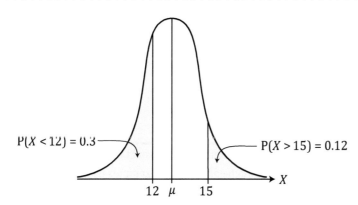

$P(X < 12) = 0.3$ $P(X > 15) = 0.12$

12 μ 15

$Z \sim N(0, 1)$ as the standard normal distribution is being used.

Looking at the higher tail, we have:

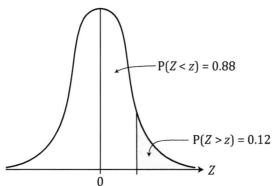

$P(Z < z) = 0.88$

$P(Z > z) = 0.12$

0

Use 'Inverse CD' on the calculator with $\mu = 0$, $\sigma = 1$, and $P(Z < z) = $ area $= 0.88$

The calculator gives a z-value of 1.1750

When $x = 15$, $$z = \frac{x - \mu}{\sigma} = \frac{15 - \mu}{\sigma}$$

Hence $$1.1750 = \frac{15 - \mu}{\sigma}$$

$$1.1750\sigma = 15 - \mu \tag{1}$$

Looking at the lower tail, we have:

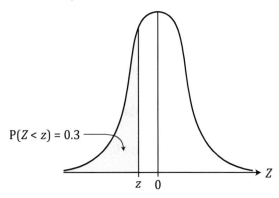

Use the 'Inverse Normal' on the calculator with $\mu = 0$, $\sigma = 1$, and $P(Z < z) = $ area $= 0.3$

Calculator gives a z-value of -0.5244

When $x = 12$, $$z = \frac{x - \mu}{\sigma} = \frac{12 - \mu}{\sigma}$$

Hence $$-0.5244 = \frac{12 - \mu}{\sigma}$$

$$-0.5244\,\sigma = 12 - \mu \tag{2}$$

Solving equations (1) and (2) simultaneously

$$\mu = 12.926 \text{ (3 d.p.)}$$

$$\sigma = 1.765 \text{ (3 d.p.)}$$

Active Learning

There are many examples in this topic where we have used tables to find answers to questions. It would be a good idea to work through each of these examples using a calculator rather than tables.

2.5 Selecting an appropriate distribution for a particular context

In this topic we have looked at two distributions for modelling continuous data: the continuous uniform distribution and the normal distribution. If you have continuous data then one of these distributions can be used.

If, however, a question is concerned with discrete data, it is necessary to use one of the probability distributions that uses discrete data such as the discrete uniform distribution, the binomial distribution or the Poisson distribution.

Many examination questions will not specify which probability distribution to use when answering a question, so you have to decide which one to use.

Here is a summary to help you to decide which distribution to use.

Here are the probability distributions you need to cover:

- Discrete uniform probability distribution
- Binomial probability distribution
- Poisson probability distribution
- Continuous uniform probability distribution
- Normal probability distribution

The first question you should ask is: 'are the data discrete or variable'. You can then choose which one of these lists to use:

> Remember that all the discrete distributions were covered in detail in the AS Applied book.

Discrete distributions

Discrete uniform distribution is used when:

- The distribution consists of discrete data.
- There is a fixed number of values.
- Each value of the discrete random variable is equally likely to occur (i.e. the probability of each value occurring is the same).

Binomial distribution is used when:

- The distribution consists of discrete data.
- There are a fixed number of trials (as denoted in the formula as n).
- There is an exact probability of the event occurring.
- Each trial is independent. (i.e. each trial should not have an effect on any other trial).
- There are only two outcomes (i.e. success and failure) to each trial.
- The probability of success is constant (i.e. there is a constant value for p, the letter used for the probability of success or failure).

Poisson distribution is used when:

- The distribution consists of discrete data.
- There are a very large number of trials (usually $n > 50$).
- The trials occur averaged out at a fixed rate over an interval of space or time. The average rate at which the events occur, λ, is known or can be found.
- Each trial is random and independent.
- Where the probability p is very small (p is usually < 0.1).
- There is success or failure for each trial.

Continuous distributions

Continuous uniform distribution is used when:

- The distribution consists of continuous data.
- Any intervals of equal length are equally likely to occur between a minimum value a and a maximum value b.

Normal distribution is used when:

- The distribution consists of continuous data.

- The distribution is bell-shaped and symmetrical about the peak value corresponding to the mean.

- The mean, median and mode are almost the same value.

- Where the mean and variance or standard deviation are known.

- Almost all the data are situated within 3 standard deviations of the mean.

- Where you want to find the probability of above or below a certain value or between a range of values.

Examples

1   Choose the most appropriate probability distribution to model each of the following situations.

   (a)   The scores obtained by rolling a fair dice 50 times.

   (b)   The distances travelled to work by 120 members of staff at a school.

   (c)   The number of passenger check-ins per hour dealt with by one member of staff at the check-in at an airport.

   (d)   The successful germination of seeds when 1000 seeds are planted, and the probability of germination is 0.9 per seed.

   (e)   A distribution where the mean, median and mode are almost the same value and where the data are continuous and symmetrical about the mean.

Answer

1   (a)   These are discrete data as the values can only be integers from 1 to 6 with each number having an equal probability of occurring. Hence use the discrete uniform probability distribution.

   (b)   Distances are continuous data. It is likely that the data will be symmetric either side of the mean, so we can use a normal distribution.

   (c)   Numbers of passengers are discrete data and we are interested in the number of events occurring randomly within a given time interval. Use a Poisson distribution.

   (d)   The number of seeds germinating are discrete data and the probability of a seed germinating is fixed and independent. The binomial distribution can be used here.

   (e)   The data are continuous and the mean, mode and median are almost the same so a normal distribution is an appropriate distribution for the model.

2   A ribbon of length 50 cm is cut at a random point.

   (a)   Name a distribution, including parameters, that can be used to model the length of the longer piece of ribbon and find its mean and variance.

   (b)   The longer piece of ribbon is shaped to form the perimeter of a circle. Find the probability that the area of the circle is greater than 100 cm^2.

Answer

2 (a) A continuous uniform distribution with parameters [25, 50]

Mean, $\mu = \dfrac{\Sigma x}{n}$

$= \dfrac{25 + 50}{2}$

$= 37.5$

$= 38$ (2 s.f.)

Variance $= \dfrac{1}{12}\left(b - a\right)^2$

$= \dfrac{1}{12}\left(50 - 25\right)^2$

$= 52.1$

> You would use a continuous uniform distribution because you are interested in the length values of the larger piece of ribbon depending on where it is cut so the length is uncountable because we don't know where it is being cut.

> In order for the piece of ribbon to be longer it must be greater than half of the ribbon's length which is 25 cm to ensure that it is the bigger piece and that it isn't the same length as the remaining piece of ribbon. The larger piece of ribbon must also be smaller than the total ribbon length of 50 cm as the ribbon is being cut so it cannot be 50 cm. Therefore the parameters are [25,50].

(b) You need to find the probability that the area of the circle is greater than 100 cm², so $P(\pi R^2 > 100)$.

Rearranging for the variable R we obtain:

$$P\left(R > \sqrt{\tfrac{100}{\pi}}\right)$$

Circumference of circle, $L = 2\pi R$, so $R = \dfrac{L}{2\pi}$

Hence, $$P\left(L > 2\pi \sqrt{\tfrac{100}{\pi}}\right)$$

Working out the numbers on a calculator, we obtain:

$$P(L > 35.45)$$

Using the limits found in part (a) to find the probability we obtain:

$$P(L > 35.45) = \dfrac{50 - 35.45}{25}$$

$$= 0.582$$

$$= 0.58 \text{ (2 s.f.)}$$

Test yourself

1. The sides of a square are of length L cm and its area is A cm^2. Given that A is uniformly distributed on the interval [15, 20], find P($L \leq 4$). [3]

2. The random variable X is normally distributed with mean 10 and standard deviation 2.
   (a) Evaluate P($X \leq 10.5$). [2]
   (b) Given that P($X \geq x$) = 0.1, find the value of x. [2]

3. X is a random variable that can be modelled by a continuous uniform distribution over the interval [1.5, 7.5].
   (a) Find the mean, variance and standard deviation of X. [3]
   (b) Find P($X < 5$). [2]

4. If $Z \sim$ N(0, 1), find P($-0.5 < z < 0.65$) [3]

5. The times taken by a group of students to complete a task are normally distributed, with a mean of 120 minutes and a standard deviation of 25 minutes.
   Find the probability that a student picked at random:
   (a) completes the task in under 130 minutes, [2]
   (b) completes the task in over 100 minutes, [2]
   (c) completes the task in between 90 and 130 minutes. [2]

6. Bags of coffee are filled with coffee beans by a machine. The weight of each bag can be modelled by a normal distribution with mean 240 g and standard deviation 15 g.
   Determine the probability that a randomly selected bag of coffee weighs:
   (a) less than 227 g [3]
   (b) between 227 g and 240 g. [3]

7. A radar unit on a motorway measures the speeds of passing vehicles. The speeds of vehicles are normally distributed with a mean speed of 65 miles per hour and a standard deviation of 15 miles per hour. What is the probability that a vehicle picked at random travelling along the motorway is travelling more than 90 miles per hour? [4]

8. The time taken, X minutes, to service a car at a garage can be modelled as a random variable using a normal distribution with mean μ and standard deviation σ.
   If P($X < 170$) = 0.14 and P($X > 200$) = 0.03, find the values of μ and σ. [6]

9. For a certain type of mobile phone, the length of time between charges of the battery is normally distributed with a mean time of 30 hours and standard deviation of 5 hours. Find the probability that the length of time between charges is between 35 and 45 hours. [3]

10. A pet food company sells bags of dog food and the weights are normally distributed with a mean of 13 kg and a standard deviation of 0.8 kg.
    (a) **Using a calculator**, find the probability that a randomly selected bag has a weight of less than 12.5 kg. [2]
    (b) Two bags are chosen at random. Find the probability that both bags will have a weight of less than 12.5 kg. Give your answer to 2 significant figures. [2]

Summary

Check you know the following facts:

Continuous uniform distribution

If $X \sim \text{U}[a, b]$, then:

$$\text{Mean, } E(X) = \frac{1}{2}(a + b)$$

$$\text{Variance, } \text{Var}(X) = \frac{1}{12}(b - a)^2$$

$$P(c \leq X \leq d) = \frac{d - c}{b - a}$$

> Both of these formulae are included in the formula booklet.

> This formula is not included in the formula booklet and will need to be remembered.

Normal distribution

If $X \sim \text{N}(\mu, \sigma^2)$ the distribution is:

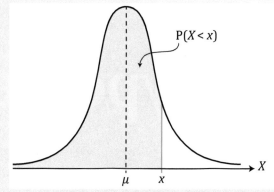

For the standard normal distribution, the distribution is adjusted to

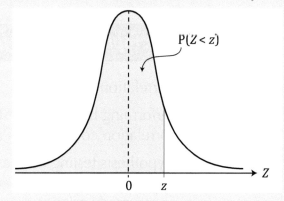

Where $z = \dfrac{x - \mu}{\sigma}$

The normal distribution function table is used to find $P(Z < z)$

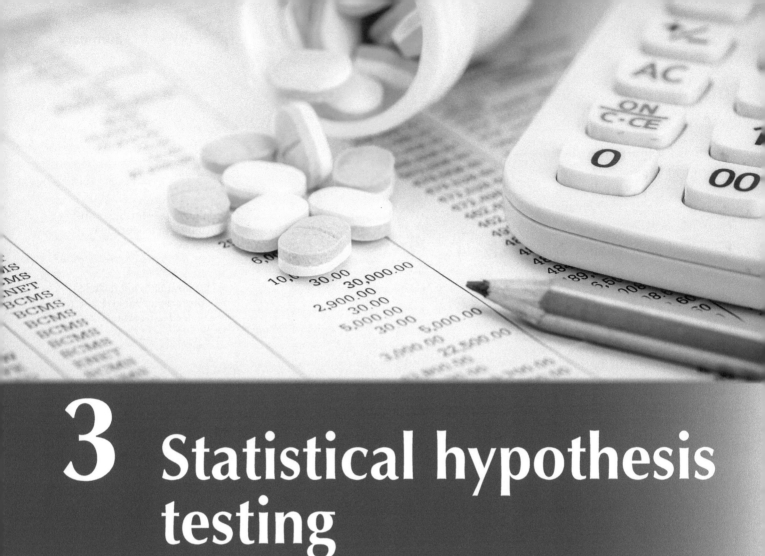

3 Statistical hypothesis testing

Introduction

Topic 5 of the AS covered statistical hypothesis testing and you need to familiarise yourself with this topic before starting this new A2 topic.

In this topic you will be introduced to a statistic called the 'product moment correlation coefficient', which is used to measure the strength of correlation between two variables where a relationship is thought to exist.

This statistic can be used as the test statistic to conduct a hypothesis test at a certain level of significance.

You will also look at how the mean of a normal distribution can be used as the test statistic when performing hypothesis testing.

3.1 Correlation coefficients

A correlation coefficient is a statistical measure that calculates the strength of the relationship between the relative movements of two variables. If a scatter diagram is drawn, the correlation coefficient measures how near the points lie to a straight line. In some cases there will be no correlation between the two variables so no straight line can be drawn and the correlation coefficient in this case is zero.

As well as the correlation coefficient measuring of the amount of scatter between two variables it also indicates the direction of the correlation (i.e. positive or negative correlation).

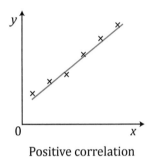
Positive correlation

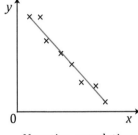
Negative correlation

There are a number of different letters given to correlation coefficients:

- The Greek letter ρ (rho) is used for the correlation coefficient for a population.

- The letter r is used for the correlation coefficient for a sample.

Just to add to the confusion, sometimes the product moment correlation coefficient is called the 'Pearson correlation coefficient', or it is given its full name, the 'Pearson product moment correlation coefficient'. In this topic we treat all coefficients in exactly the same way, so don't worry about the name given.

A correlation coefficient can take any value between and including −1 and 1.

A positive correlation coefficient shows positive correlation and the nearer its value is to 1, the stronger the correlation and the nearer the scatter graph will be to points lying in a straight line.

If there is perfect negative correlation, then the correlation coefficient will equal −1.

If the correlation coefficient equals zero, then this indicates no correlation, maybe because there is no linear relationship between the variables.

The scatter diagrams below show the various types of correlation between two variables x and y and the corresponding values for the correlation coefficient.

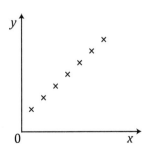
Perfect positive correlation, correlation coefficient = 1.

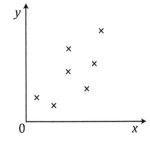
Low positive correlation, correlation coefficient = 0.5 .

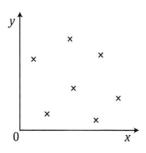

No correlation,
correlation coefficient = 0.

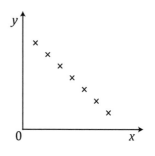

Perfect negative correlation,
correlation coefficient = –1.

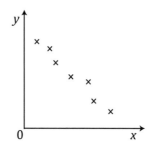

High negative correlation,
correlation coefficient = –0.9

The Pearson correlation coefficient is also called the product moment correlation coefficient (PMCC)

Remember that the letter you use for a correlation coefficient depends on whether a population or a sample was used to calculate it. You do not have to calculate the value of a correlation coefficient from a set of data. You only have to be able to use it.

The Pearson correlation coefficient for the population is denoted by the Greek letter ρ and can take any value between and including –1 and 1.

> All correlation coefficients can take values between and including –1 and 1

The Pearson correlation coefficient for a sample is denoted by the letter r and can take any value between and including –1 and 1.

Using the correlation coefficient to measure correlation

The correlation coefficient can be used to interpret the degree with which two quantities are correlated in a particular context as the following example shows.

Example

1   Amy conducts an investigation to see if there is correlation between the latitude north of the equator and the mean temperature. After studying and processing weather data at certain latitudes north of the equator she calculates the product moment correlation coefficient to be –0.875. Interpret this result in context.

> It is important to note that when you interpret your result in context, you should always say that the data 'suggest ...'. Remember that the result is based on the data collected, so it is impossible to be completely sure.

Answer

1   The value –0.875 is very close to –1 and this means that there is strong negative correlation and the data suggest that as the latitude north of the equator increases, the average temperature decreases.

3.2 Performing hypothesis testing using a correlation coefficient as a test statistic

A test statistic is a statistic that is used for hypothesis testing. The test statistic used must be a statistic that is a single value that can be used to represent the data set.

A correlation coefficient can be used as a test statistic for hypothesis testing.

Suppose we wanted to test whether two variables are correlated or not. We can use the correlation coefficient to conduct either a one- or two-tailed test.

A one-tailed test can be used to determine whether the PMCC, denoted by the Greek letter ρ, is greater than zero using the following hypothesis test:

$$\mathbf{H}_0 : \rho = 0, \ \mathbf{H}_1 : \rho > 0$$

Notice that the **null hypothesis** is that there is **no** correlation between the two variables.

If you want to determine whether the PMCC is less than zero, you can conduct the following one-tailed test:

$$\mathbf{H}_0 : \rho = 0, \ \mathbf{H}_1 : \rho < 0$$

If you want to determine whether the PMCC for a population is not equal to zero you can conduct the following two-tailed test:

$$\mathbf{H}_0 : \rho = 0, \ \mathbf{H}_1 : \rho \neq 0$$

> It is important to note that there are two correlation coefficients depending on whether they are based on a sample or an entire population. The correlation coefficient based on a sample is given the letter r, and the coefficient based on the population is given the Greek letter rho (i.e. ρ)

Critical values for hypothesis testing

In hypothesis testing, a critical value is a point on the test distribution that is compared to the test statistic to determine whether to reject the null hypothesis. If the absolute value of your test statistic is greater than the critical value, you can declare statistical significance and reject the null hypothesis.

The steps you need to take to perform a hypothesis test using the correlation coefficient as the test statistic

1   Write the null hypothesis in terms of the correlation coefficient.

$$\mathbf{H}_0 : \rho = 0$$

2   Decide from the wording of the question whether you are investigating positive correlation, negative correlation or any correlation (i.e. positive or negative).

3   Write the alternative hypothesis in terms of the correlation coefficient (e.g. $\mathbf{H}_1 : \rho > 0$, $\mathbf{H}_1 : \rho < 0$ or $\mathbf{H}_1 : \rho \neq 0$).

Note that for $\mathbf{H}_1 : \rho > 0$ or $\mathbf{H}_1 : \rho < 0$ these are both one-tailed tests. For $\mathbf{H}_1 : \rho \neq 0$ this will be a two-tailed test.

4   Obtain a copy of the statistical tables and look at Table 9 'Critical values of the product moment correlation coefficient'.

Look at the top and choose from one-tail or two-tail for the level of significance you are using. Note that the significance level is given as a percentage.

Now look down the first column for the number of pairs of data values, n, you are using, then read across to where the column for the significance intersects.

Where the two intersect is the critical value.

5   Compare the value of the correlation coefficient with this critical value.

If the correlation coefficient > critical value the null hypothesis is rejected, indicating evidence of correlation.

6   State the meaning of the results to the hypothesis test within the context of the question.

Now follow these examples which will enable you to see how we use these steps to answer questions on hypothesis testing using the correlation coefficient.

Examples

1   A teacher suspects that students in his class who are good at maths are also good at physics, so he collects pairs of maths and physics marks for 20 students.

The teacher uses these pairs of marks to calculate the product moment correlation coefficient and finds it to be 0.68. The teacher has decided to test for positive correlation.

(a)   State the null and alternative hypotheses the teacher could use for this test.

(b)   Using the 'Critical values of the product moment correlation coefficient' table carry out a hypothesis test at the 5% level of significance and state whether the results are significant. Write a conclusion in context.

Note that for the null hypothesis, we assume that there is no correlation, so the correlation coefficient would be zero.

The alternative hypothesis is that there is positive correlation, so the correlation coefficient is greater than 0.

Answer

1   (a)   Null hypothesis is $H_0 : \rho = 0$

Alternative hypothesis is $H_1 : \rho > 0$

(b)   This is a one-tailed test, so looking up 5% significance level for a one-tail test and $n = 20$, the critical value is read off from the 'Critical values of the product moment correlation coefficient' table.

TABLE 9 CRITICAL VALUES OF THE PRODUCT MOMENT CORRELATION COEFFICIENT

The table gives the critical values, for different significance levels, of the sample product moment correlation coefficient r based on n independent pairs of observations from a bivariate normal distribution with correlation coefficient $\rho = 0$.

One tail	10%	5%	2.5%	1%	0.5%
Two tail	20%	10%	5%	2%	1%
n					
4	0.8000	0.9000	0.9500	0.9800	0.9900
5	0.6870	0.8054	0.8783	0.9343	0.9587
6	0.6084	0.7293	0.8114	0.8822	0.9172
7	0.5509	0.6694	0.7545	0.8329	0.8745
8	0.5067	0.6215	0.7067	0.7887	0.8343
9	0.4716	0.5822	0.6664	0.7498	0.7977
10	0.4428	0.5494	0.6319	0.7155	0.7646
11	0.4187	0.5214	0.6021	0.6851	0.7348
12	0.3981	0.4973	0.5760	0.6581	0.7079
13	0.3802	0.4762	0.5529	0.6339	0.6835
14	0.3646	0.4575	0.5324	0.6120	0.6614
15	0.3507	0.4409	0.5140	0.5923	0.6411
16	0.3383	0.4259	0.4973	0.5742	0.6226
17	0.3271	0.4124	0.4821	0.5577	0.6055
18	0.3170	0.4000	0.4683	0.5425	0.5897
19	0.3077	0.3887	0.4555	0.5285	0.5751
20	0.2992	0.3783	0.4438	0.5155	0.5614
21	0.2914	0.3687	0.4329	0.5034	0.5487

The critical value is shown highlighted in the above table.

Critical value = 0.3783

The critical value is compared against the test statistic, which is the product moment correlation coefficient.

We would reject the null hypothesis H_0 if the product moment correlation coefficient was greater than the critical value (i.e. 0.3783).

Now the PMCC = 0.68 and as 0.68 > 0.3783 we can say the result is significant.

Hence at the 5% level of significance there is evidence to reject the null hypothesis in favour of the alternative hypothesis that the marks in physics and maths are positively correlated.

BOOST

Grade ⬆⬆⬆⬆

Always say there is evidence to reject/accept the null hypothesis. Never say that the null hypothesis or alternative hypothesis should be simply 'accepted'.

2   Two variables X and Y that are thought to correlated in some way. The population of 20 pairs of values was obtained and the product moment correlation coefficient was found to be 0.75.

(a)   State two hypotheses that can be used to test whether X and Y are correlated.

(b)   Test, using a 5% significance level, whether this correlation is significant and state your conclusion in context.

- -

Answer

2   (a)   If ρ is the product moment correlation coefficient then the two hypotheses to be considered are:

$H_0 : \rho = 0$

$H_1 : \rho \neq 0$

The null hypothesis is that there is no correlation in which case the PMCC would be zero.

Note that as we do not know whether the X and Y are positively or negatively correlated, this is a two-tailed test.

(b)   PMCC = 0.75 and $n = 20$.

Because we are looking at $H_1 : \rho \neq 0$ we need to consider $\rho < 0$ and $\rho > 0$ so this test is a two-tailed test.

We now find the critical value of the product moment correlation coefficient by looking it up in Table 9 of Elementary Statistical Tables.

We are conducting a two-tailed test at a 5% significance level, so we look along the column to the 5% two-tail test. We then look for a value of n equal to 20 (i.e. the number of pairs of observations) and then look for the intersection of the column and row. You can see the critical value is ±0.4438 . (Note that we include the ± as we are looking at both tails of the normal distribution curve as this is a two-tailed test.)

The product moment correlation coefficient is used as the test statistic and this is compared with the critical value obtained from tables to see if the null hypothesis should be rejected or not.

Now the test statistic the PMCC, 0.75 > the critical value, 0.4438

This result is significant.

This means that there is evidence to reject the null hypothesis and that the product moment correlation coefficient is greater than 0 and that there is evidence at the 5% level of significance that there is correlation between the variables X and Y.

BOOST

Grade ⬆⬆⬆⬆

Notice that the question asks whether the result is significant or not. You must make sure you supply an answer to this.

3   A student in Year 13 decides to see if there is a relationship for students in her year between shoe size (X) and height (Y) so she collects the pairs of data using random sample selected from the population which is all the students in Year 13 and the following results are obtained:

Shoe size (X)	Height (inches) (Y)
3	58
7.5	64
6	65
8	70
7	63
8.5	69
9	71
11	76
12.5	73
9	70
4	61
5	63
8	69
4	59
6	60
10	73
9	64
8.5	70

The value of the product moment correlation coefficient calculated from this data is found to be 0.9063.

(a)   A hypothesis test is to be performed using the correlation coefficient, ρ, as the test statistic.

   (i)   State, giving a reason, whether a one-tailed or a two-tailed test should be used.

   (ii)   Write down suitable null and alternative hypotheses.

(b)   Test at the 5% level of significance, whether there is evidence of a correlation between the height of a person and their shoe size in this group of students.

· ·

Answer

3   (a)   (i)   A two-tailed test as we need to test for positive as well as negative correlation.

   (ii)   Null hypothesis $H_0 : \rho = 0$

   Alternative hypothesis is $H_1 : \rho \neq 0$

The sample size is found by counting the number of pairs of values.

(b)   The sample size = 18

The test statistic is the product moment correlation coefficient = 0.9063.

We now use Table 9 in the Elementary Statistical Tables to determine the critical values for the product moment correlation coefficient. We are conducting a two-tailed test at a 5% significance level so we look along the column to the 5% two-tail test. We then look for a value of n equal to 18 (i.e. the number of pairs of observations) and then look for the intersection of the column and row. You can see the critical values are ±0.4683. Notice that as this is a two-tailed test we include both the plus and minus signs.

Since the test statistic 0.9063 > 0.4683, H_0 is rejected.

There is strong evidence to suggest that the correlation coefficient is greater than zero and that there is positive correlation between shoe size and height.

Step by STEP

A hotel owner in Cardiff is interested in what factors hotel guests think are important when staying at a hotel. From a hotel booking website she collects the ratings for 'Cleanliness', 'Location' and 'Value for money' for a random sample of 17 Cardiff hotels.

(Each rating is an average of all scores awarded by guests who have contributed reviews using a scale from 1 to 10, where 10 is 'Excellent'.)

The scatter graph below shows the relationship between 'Value for money' and 'Cleanliness' for the sample of Cardiff hotels.

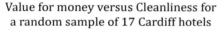

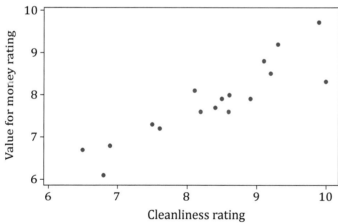

(a) The product moment correlation coefficient for 'Value for money' and 'Cleanliness' for the sample of 17 Cardiff hotels is 0.895.

Stating your hypothesis clearly, test, at the 5% level of significance, whether this correlation is significant. State your conclusion in context.

(b) The hotel owner also wishes to investigate whether 'Value for money' has a significant correlation with 'Cost per night'. She used a statistical analysis package which provided the following output which includes the Pearson correlation coefficient of interest and the corresponding p-value which is bracketed in the table.

	Value for money	Cost per night
Value for money	1	
Cost per night	0.047 (0.859)	1

Comment on the correlation between 'Value for money' and 'Cost per night'.

Steps to take

1   In part (a) we use the PMCC of 0.895 as a test statistic at a 5% significance level to determine the critical value which is then compared with the test statistic.

We need to first state the null hypothesis that there is no correlation so the PMCC will be zero. The alternative hypothesis is that there is correlation and that the PMCC does not equal zero.

2   Use Table 9 critical values of the product moment correlation coefficient and look up a two-tailed test at 5% significance level with $n = 17$.

3   Compare the critical values from the table with 0.895.

4   If the PMCC (i.e. the test statistic) is greater than the critical value, then the null hypothesis should be rejected, otherwise it should be kept.

5   Add a comment, making clear the context of your answer.

6   In part (b) notice that there is no significance level mentioned and you do not know the value of n.

Here we use the p-value given and compare it with the following p-values:

$p < 0.01$; there is very strong evidence for rejecting $\mathbf{H}_0$,

$0.01 \leq p \leq 0.05$; there is strong evidence for rejecting $\mathbf{H}_0$,

$p > 0.05$; there is insufficient evidence for rejecting $\mathbf{H}_0$.

The correlation coefficient can be compared, so see how near it is to 1 or −1 or to zero in order to assess the correlation.

7   A comment needs to then be made in the context of the question as to whether they are correlated or not making reference to the evidence.

Don't get mixed up between the correlation coefficient ρ and the p-value. The Greek letter rho (ρ) looks very similar to an italicised p.

. . . .

Answer

(a)   If ρ is the product moment correlation coefficient then the two hypotheses to be considered are:

$$\mathbf{H}_0 : \rho = 0$$

$$\mathbf{H}_1 : \rho \neq 0$$

The test statistic is the product moment correlation coefficient = 0.895

We now use Table 9 in the Elementary Statistical Tables to determine the critical values for the product moment correlation coefficient. We are conducting a two-tailed test at a 5% significance level so we look along the column to the 5% two-tail test. We then look for a value of n equal to 17 (i.e. the number of pairs of observations) and then look for the intersection of the column and row. You can see the critical value is ±0.4821

TABLE 9 CRITICAL VALUES OF THE PRODUCT MOMENT CORRELATION COEFFICIENT

The table gives the critical values, for different significance levels, of the sample product moment correlation coefficient r based on n independent pairs of observations from a bivariate normal distribution with correlation coefficient $\rho = 0$.

One tail Two tail n	10% 20%	5% 10%	2.5% 5%	1% 2%	0.5% 1%
4	0.8000	0.9000	0.9500	0.9800	0.9900
5	0.6870	0.8054	0.8783	0.9343	0.9587
6	0.6084	0.7293	0.8114	0.8822	0.9172
7	0.5509	0.6694	0.7545	0.8329	0.8745
8	0.5067	0.6215	0.7067	0.7887	0.8343
9	0.4716	0.5822	0.6664	0.7498	0.7977
10	0.4428	0.5494	0.6319	0.7155	0.7646
11	0.4187	0.5214	0.6021	0.6851	0.7348
12	0.3981	0.4973	0.5760	0.6581	0.7079
13	0.3802	0.4762	0.5529	0.6339	0.6835
14	0.3646	0.4575	0.5324	0.6120	0.6614
15	0.3507	0.4409	0.5140	0.5923	0.6411
16	0.3383	0.4259	0.4973	0.5742	0.6226
17	0.3271	0.4124	0.4821	0.5577	0.6055

Now the test statistic, 0.895 > the critical value, 0.4821

This means there is sufficient evidence to reject the null hypothesis and that the product moment correlation coefficient is not equal to 0 and that there is evidence at the 5% level of significance that there is correlation.

(b) The p-value for correlation between 'Value for money' and 'Cost per night' (i.e. 0.859) is > 0.05. As the p-value is > 0.05 the evidence suggests the two quantities are not correlated.

The Pearson correlation coefficient of 0.047 is quite near to zero, which also indicates no correlation is likely.

Hence we can conclude that 'Cost per night' does not seem to be correlated to 'Value for money'.

3.3 Hypothesis testing for the mean of a normal distribution with a known, given or assumed variance

In this section, we will be looking at conducting a hypothesis test for the mean of a normal distribution with known, given or assumed variance. Suppose lots of samples of size n of a random variable X were taken from a population, then if the population is normally distributed the means of the samples $\overline{X}$ will also be normally distributed.

If you have a random variable X and it is normally distributed with a population mean μ and variance σ^2 then you can test hypotheses about the population mean by looking at the mean, $\overline{x}$, of the sample.

If X is normally distributed then we can say $X \sim N(\mu, \sigma^2)$ then the sample mean, $\overline{X}$, is normally distributed and we can say, $\overline{X} \sim N\left(\mu, \frac{\sigma^2}{n}\right)$

If $Z \sim N(0, 1)$, then $$Z = \frac{\overline{X} - \mu}{\frac{\sigma}{\sqrt{n}}}$$

This formula is used to work out z-values that apply to the standard normal distribution.

Hence the z-value (i.e. the test statistic) will be

$$z = \frac{\overline{x} - \mu}{\frac{\sigma}{\sqrt{n}}}$$

The z-value can then be used to find critical values and critical regions that can be used for hypothesis testing.

Using a significance level to find a critical value and hence perform a one-tailed hypothesis test

In the following example, you are told the significance level and can then use this to find the z-value and hence the critical value. Once the critical value has been found, you can then perform a hypothesis test to see if the sample mean lies inside or outside the critical region. If it lies inside the critical region, the result is significant and there is evidence that the null hypothesis should be rejected.

The following example shows this method for a one-tailed test.

Example

1   A machine in a factory fills bags of coffee. It has been found that the bags have a population mean weight of 235 grams and a standard deviation of 15 grams. It has been established that the weight of bags can be modelled using a normal distribution.

The production manager is worried that the machine filling the beans is malfunctioning and it is putting less coffee into the bags. She takes a sample of 20 bags to see if the mean weight of bags is different from the population mean of 235 grams.

(a)   Assuming that the weights of bags in the sample is normally distributed and still with a standard deviation of 15 grams and using a 1% significance level, determine the critical value for the weight of bags.

(b)   The production manager takes a sample of bags and calculates the mean from her sample. The mean weight of her sample was 226.5 grams.

State what conclusion the production manager can draw from this based on your answer to part (a).

. . . .

Answer

1   (a)   The null hypothesis is that the mean weight of coffee in the sample is equal to 235 g.

The alternative hypothesis is that the mean weight of coffee in the sample is less than 235 g

> Note that as we are only investigating the lower tail, this is a one-tailed test.

Hence we can write: $\mathbf{H_0} : \mu = 235$ $\mathbf{H_1} : \mu < 235$

Let X be the weight of coffee in a bag.

Assuming $\mathbf{H_0}$, if $X \sim \mathrm{N}(\mu, \sigma^2)$.

We now use Table 3 Normal Distribution Function table to determine the critical z-value.

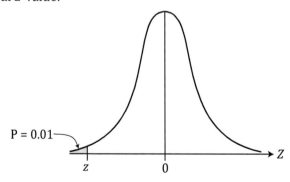

We are conducting a one-tailed test at a 1% significance level, so we look in the body of the table for a probability as near to (1 − 0.01 = 0.99) as possible. We can use Table 4 to look up the z-value for the probability of 0.99. From the table a z-value of 2.326 is obtained, which is changed to −2.326 as from the sketch we know that the z-value is negative.

This z-value is then used to find the corresponding value of x which is then the critical value.

$$z = \frac{\bar{x} - \mu}{\frac{\sigma}{\sqrt{n}}}$$

$$-2.326 = \frac{x - 235}{\frac{15}{\sqrt{20}}}$$

Giving critical value, $\bar{x} = 227.2$

(b)   The sample mean of 226.5 grams is below the critical value of 227.2 and it therefore lies in the critical region. We can say that this result is significant.

This means there is sufficient evidence at the 1% level of significance, that the machine is putting less coffee into the bags and is therefore malfunctioning.

Hence there is evidence that the production manager is right in that the machine is malfunctioning and putting less coffee in the bags.

> It is always best to draw a sketch of the standard normal distribution, so you know on which side of the distribution the critical value lies. Here we are testing for a lower value, so the critical value lies in the lower tail.

> **BOOST**
> Grade ⇧⇧⇧⇧
> Do not simply say that the null hypothesis is rejected in favour of the alternative hypothesis. Your answer needs to be given in the context of the question.

Using a significance level to find critical values and hence perform a two-tailed hypothesis test

If you want to investigate a value that has changed but you don't know whether it has got larger or smaller, you need to conduct a two-tailed test using the mean of the sample.

The following example shows the method.

Example

1   A machine is filling bottles with a volume V of shampoo where V is normally distributed with a mean of 105 ml and standard deviation of 0.75 ml.

A new part has been fitted to the machine and it is suspected that the volume it is filling the bottles with has changed. A sample of 30 bottles of shampoo was taken and the volume measured accurately to see if the mean volume had changed from 105 ml. The volume of bottles in the sample was still found to be normally distributed with the same standard deviation (i.e. 0.75 ml) as before.

(a)   State two hypotheses that could be used to determine if the volume had changed.

(b)   Conduct a significance test at the 1% level of significance to determine the critical regions for the test.

The mean volume of the sample of 30 bottles was calculated to be 104 ml.

(c)   What conclusion can be drawn about the volumes of bottles after the new part was fitted?

Answer

This is a two-tailed test to see if the volume is higher or lower than 105 ml.

1   (a) $\mathbf{H}_0 : \mu = 105$

$\mathbf{H}_1 : \mu \neq 105$

(b)   Assuming the null hypothesis, $\mathbf{H}_0$, we have $V \sim N(105, 0.75^2)$.

The mean value of the sample is also assumed to be normally distributed, so we have:

$$\bar{V} \sim N\left(105, \frac{0.75^2}{30}\right)$$

Now we standardise the normal distribution by finding the z-value.

The z-value now needs to be found using tables so it can be substituted back into this equation so one of the critical values can be found. The other z-value will have the same value but opposite sign.

Now

$$z = \frac{\bar{V} - \mu}{\frac{\sigma}{\sqrt{n}}}$$

$$z = \frac{\bar{V} - 105}{\frac{0.75}{\sqrt{30}}}$$

As the significance level is 1% and we are performing a two-tailed test we need to halve this value so the probability/area in each tail is 0.005.

We now need to find the z-values by using Table 4 Percentage Points of the Normal Distribution and looking for a z-value corresponding to a probability of $1 - 0.005 = 0.995$. From the table, a probability of 0.995 gives a z-value of 2.576.

As this is a two-tailed test there are two z-values, -2.576 and 2.576. These are shown on the standard normal distribution below:

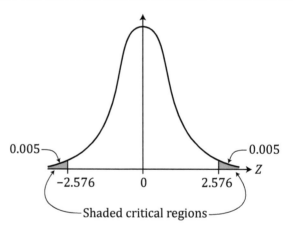

Shaded critical regions

We can now use the formula $z = \frac{\bar{V} - 105}{\frac{0.75}{\sqrt{30}}}$

Here we use the formula for the z-value in reverse to find the critical values for the volumes.

to find the critical values for the volume, so that the critical region can be determined.

$$2.576 = \frac{\bar{V} - 105}{\frac{0.75}{\sqrt{30}}} \text{ and this gives } \bar{v} = 105.3527$$

$$-2.576 = \frac{\bar{V} - 105}{\frac{0.75}{\sqrt{30}}} \text{ and this gives } \bar{v} = 104.6473$$

This means the critical regions are $\overline{V} \leq 104.6473$ or $\overline{V} \geq 105.3527$

(c) The mean from the sample is 104 ml and this is inside the critical region meaning there is significant evidence at the 1% significance level that the mean volume has changed due to the new part.

Solution using a calculator

Here we will look at the method using a calculator to answer part (b) of this question.

First use the calculator to work out $\frac{\sigma}{\sqrt{n}}$ i.e. $\frac{\sigma}{\sqrt{n}} = \frac{0.75}{\sqrt{30}} = 0.1369$

As this is a two-tailed test, we need to find the critical values in the upper and lower tails.

For the upper tail:

Set your calculator for 'Inverse Normal' and enter the following parameters:

- Enter the area $(1 - 0.005) = 0.995$
- Enter $\sigma = 0.1369$
- Enter $\mu = 105$

Answer, $\overline{V} = 105.3526$ which is the critical value in the upper tail.

> Note that when entering the value for σ you should enter the value for $\frac{\sigma}{\sqrt{n}}$

For the lower tail:

Set your calculator for Inverse Normal and enter the following parameters:

- Enter the area $= 0.005$
- Enter $\sigma = 0.1369$
- Enter $\mu = 105$

Answer, $\overline{V} = 104.6474$ which is the critical value in the lower tail.

Hence, the critical regions are $\overline{V} \leq 104.647$ or $\overline{V} \geq 105.353$

The mean from the sample is 104 ml and this is inside the critical region meaning there is significant evidence at the 1% significance level that the mean volume has changed due to the new part.

Using a *p*-value to conduct a hypothesis test

In the previous section you learned how to conduct a hypothesis test by finding the critical area and/or the critical region. The test statistic of the sample mean was then used to see if it was inside or outside the critical region to decide whether to accept the null hypothesis or reject it.

In this section we will look at a different method which uses *p*-values for hypothesis testing.

p-values

p-values were covered in Topic 5 of the AS Applied book. *p*-values are used to determine how well or otherwise the sample data support the argument that the null hypothesis is true. The *p*-value is a measure of how likely the observed effect in your sample data is, if the null hypothesis is true.

The smaller the *p*-value, the stronger the sample evidence is against the null hypothesis.

BOOST

Grade

It is important to note that if $p > 0.05$ we cannot conclude that H_0 is true, just that there is not enough evidence to reject it.

High *p*-values indicate your data are likely with a true null hypothesis.

Low *p*-values indicate your data are unlikely with a true null hypothesis.

The definition of a *p*-value is: A *p*-value is the probability of obtaining a value at least as extreme as the one in your sample given the null hypothesis is true.

We generally interpret *p*-values along the following lines:

$p < 0.01$; there is very strong evidence for rejecting H_0

$0.01 \leq p \leq 0.05$; there is strong evidence for rejecting H_0.

$p > 0.05$; there is insufficient evidence for rejecting H_0.

Examples

1 The headteacher of a primary school tries out a new method of teaching tables. Using the old method the mean mark for pupils in the school was normally distributed with a mean of 14 marks and with a standard deviation of 3 marks.

The next year the new teaching method was used and a sample of 20 students' marks was used and found to have a mean of 15.5 marks. It was assumed that the standard deviation stayed the same as before.

Test at the 5% level of significance, whether there is evidence that the mean mark had increased as a result of the new teaching method.

Answer

Notice we are interested in an increase in the marks, so this is a one-tailed test.

Alternatively you can use a calculator to work out the *p*-value.

Set the calculator to 'Normal CD' and type in:

- Lower = 15.5
- Upper = 1×10^{99}
- $\sigma = \dfrac{3}{\sqrt{20}}$ and
- $\mu = 14$.

Answer = 0.0126

1   Assuming H_0, if $X \sim N(\mu, \sigma^2)$ then the sample mean mark, $\overline{X}$, is normally distributed, so we can say $\overline{X} \sim N\left(\mu, \dfrac{\sigma^2}{n}\right)$.

$$H_0 : \mu = 14$$
$$H_1 : \mu > 14$$

Test statistic, $z = \dfrac{\overline{X} - \mu}{\dfrac{\sigma}{\sqrt{n}}}$

$$z = \dfrac{15.5 - 14}{\dfrac{3}{\sqrt{20}}}$$

$$= 2.2361$$

$$P(Z > 2.24) = 1 - P(Z < 2.24)$$

$$= 1 - 0.98745$$

$$= 0.01255$$

$$p\text{-value} = 0.0126$$

Now as *p*-value (0.0126) < 0.05 (i.e. the significance level) there is evidence that the null hypothesis should be rejected.

Hence there is evidence at the 5% level of significance that the mean mark has increased due to the new teaching method.

2 A sales representative travels a mean distance of 450 miles each week visiting customers. The company she works for has introduced a new computer system that works out an order in which she visits clients each week and routes to take in order to reduce the mileage. After the first 30 weeks after the introduction of the new system the distance travelled is normally distributed with a mean distance of 435 miles and a standard deviation of 48 miles.

Perform a hypothesis test to test whether the mean distance travelled has changed from the previous figure of 450 miles per week.

Answer

2 Let X be the distance travelled in a week after the introduction of the new computer system.

Assuming $\mathbf{H}_0$ if $X \sim N(\mu, \sigma^2)$ then the sample mean distance, $\overline{X}$, is normally distributed so we can say $\overline{X} \sim N\left(\mu, \frac{\sigma^2}{n}\right)$.

$$\mathbf{H}_0 : \mu = 450$$

$$\mathbf{H}_1 : \mu \neq 450$$

$$\overline{X} \sim N\left(450, \frac{48^2}{n}\right)$$

Test statistic, $z = \dfrac{\overline{X} - \mu}{\frac{\sigma}{\sqrt{n}}}$

$$z = \dfrac{435 - 450}{\frac{48}{\sqrt{30}}}$$

$$= -1.7116$$

$$P(Z < -1.71) = 1 - P(Z < 1.71)$$

$$= 1 - 0.95637 = 0.04363$$

$$p\text{-value} = 2 \times 0.04363 = 0.08726 = 0.087 \text{ (2 s.f.)}$$

> Alternatively you can use a calculator to work out the p-value.
>
> Set the calculator to 'Normal CD' and type in:
>
> - Lower = 1×10^{-99}
> - Upper = 435
> - $\sigma = \dfrac{48}{\sqrt{30}} = 8.76356$
> - $\mu = 450$.
>
> Answer = 0.04348
>
> As we have only used the lower tail, the probability displayed by the calculator needs to be doubled to give the p-value.
>
> So p-value = 0.08696 = 0.087 (2 s.f.)

> Notice we are interested in a change in the distance so this is a two-tailed test.

As p-value (i.e. 0.087) > 0.05, there is insufficient evidence for rejecting $\mathbf{H}_0$. Hence we can conclude that there is evidence that the new computer software has not changed the mean distance travelled per week.

> As this is a two-tailed test we need to double the probability to give the p-value.

Step by STEP

Jim is a tennis player. His serve has a mean speed of 120 miles per hour (mph). He buys a new racket and he wishes to investigate whether or not using this racket changes the mean speed of his serve. The mean speed of serve can be considered as a random variable that follows a normal distribution.

He therefore goes to a tennis centre where he hits 10 serves and the measured speeds are as follows (mph):

121.2, 119.1, 118.3, 120.1, 117.9, 118.3, 119.4, 119.6, 120.3, 117.8.

You may assume that this is a random sample from a normal distribution with a standard deviation of 1.2 .

(a) State suitable hypotheses for his investigation.

(b) Determine the *p*-value of these results and state your conclusion in context.

Steps to take 👣 👣 👣 👣

1   There is no mean value given, so this needs to be calculated first before the hypotheses can be constructed.

The mean speed is found by adding all the speeds and then dividing by 10 to give the mean speed.

2   Think about whether this is a one- or a two-tailed test. As the question says, you want to know whether the speed has changed (i.e. it could have increased or decreased), so this is a two-tailed test.

3   Write the hypothesis using the usual convention with the null hypothesis that the mean speed has not changed and the alternative hypothesis is that the mean speed has changed.

4   Enter the sample mean, standard deviation, and number of values of speed in the sample into the formula $Z = \dfrac{\overline{X} - \mu}{\frac{\sigma}{\sqrt{n}}}$ to find the *z*-value.

5   Use the normal distribution table, look for the *z*-value and read off the corresponding *p*-value.

6   Compare the *p*-value of the sample with the following *p*-values:

$$p < 0.01; \text{ there is very strong evidence for rejecting } \mathbf{H_0},$$

$$0.01 \leq p \leq 0.05; \text{ there is strong evidence for rejecting } \mathbf{H_0},$$

$$p > 0.05; \text{ there is insufficient evidence for rejecting } \mathbf{H_0}.$$

7   State the conclusion in context.

. . . .

Answer

(a) $\mathbf{H_0} : \mu = 120$

$\mathbf{H_1} : \mu \neq 120$

(b)   Let *X* be the mean speed of a tennis ball which is normally distributed.

Assuming $\mathbf{H_0}$, if $X \sim N(\mu, \sigma^2)$ then the sample mean speed, $\overline{X}$, is normally distributed so we can say $\overline{X} \sim N\left(\mu, \dfrac{\sigma^2}{n}\right)$.

Sample mean, $\overline{x} = \dfrac{\Sigma x}{n}$

$$\overline{x} = \frac{\Sigma x}{n} = \frac{121.2 + 119.1 + 118.3 + 120.1 + 117.9 + 118.3 + 119.4 + 119.6 + 120.3 + 117.8}{10}$$

$$= 119.2$$

Test statistic, $z = \dfrac{\overline{X} - \mu}{\frac{\sigma}{\sqrt{n}}}$

$$= \frac{119.2 - 120}{\frac{1.2}{\sqrt{10}}}$$

$$= -2.11$$

Note that here we are looking for a change in the speed. As the speed could increase or decrease we need to conduct a two-tailed test.

Active Learning

Work through the answer to this question, except this time use a calculator rather than tables. Make sure you add explanatory notes to explain your workings.

3.3 Hypothesis testing for the mean of a normal distribution with a known, given or assumed variance

We now need to work out how to work out the probability using the normal distribution table. Drawing sketches to help us, we have:

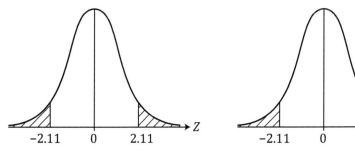

The first sketch shows the two tails shaded and the second diagram shows how to find the probability of the upper-tail using the tables.

From the Normal Distribution Function tables, a z-value of 2.11 is looked up which gives a value of 0.98257. This value is subtracted from 1 to give the p-value for the upper tail.

The other tail will have the same probability (i.e. 0.01743)

Hence the p-value for the probability = 2 × 0.01743 = 0.03486

As the p-value of 0.03486 is less than 0.05, there is strong evidence that the mean speed has changed.

Remember if you are conducting a two-tailed test you will need to double the probability (i.e. p-value) for a single tail.

We generally interpret p-values along the following lines:

$p < 0.01$; there is very strong evidence for rejecting $\mathbf{H_0}$

$0.01 \leq p \leq 0.05$; there is strong evidence for rejecting $\mathbf{H_0}$

$p > 0.05$; there is insufficient evidence for rejecting $\mathbf{H_0}$.

Example

1 Automatic coin counting machines sort, count and batch coins. A particular brand of these machines rejects 2p coins that are less than 6.12 grams or greater than 8.12 grams.

(a) The histogram represents the distribution of the weight of UK 2p coins supplied by the Royal Mint. This distribution has mean 7.12 grams and standard deviation 0.357 grams.

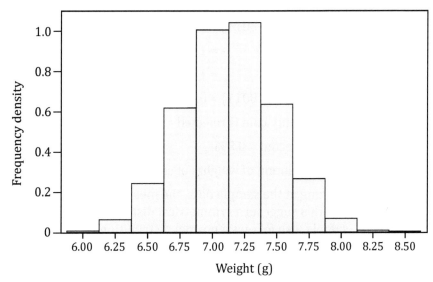

Weight of UK two pence coins

Explain why the weight of 2p coins can be modelled using a normal distribution.

(b) Assume the distribution of the weight of 2p coins is normally distributed. Calculate the proportion of 2p coins that are rejected by this brand of coin counting machine.

A manager suspects that a large batch of 2p coins is counterfeit. A random sample of 30 of the suspect coins is selected. Each of the coins in the sample is weighed. The results are shown in the summary statistics table below.

Summary statistics
Weights (in grams) for a random sample of 30 UK 2p coins

Mean	Standard Deviation	Minimum	Lower quartile	Median	Upper quartile	Maximum
6.89	0.296	6.45	6.63	6.88	7.08	7.48

(c) (i)   What assumption must be made about the weights of coins in this batch in order to conduct a test of significance on the sample mean? State, with a reason, whether you think this assumption is reasonable.

(ii) Assuming the population standard deviation is 0.357 grams, test at the 1% significance level whether the mean weight of the 2p coins in this batch is less than 7.12 grams.

. . . .

Answer

1   (a)   The graph is bell shaped (with an approximately even distribution either side of the mean).

(b)   Coins are rejected if $X < 6.12$ or $X > 8.12$. We need to find the probability of each of these and add the probabilities together.

> The coins are rejected if they are too heavy or too light.

$$z = \frac{x - \mu}{\sigma} = \frac{6.12 - 7.12}{0.357} = -2.8011$$

> We first find the z-values for each of the X values.

$$z = \frac{x - \mu}{\sigma} = \frac{8.12 - 7.12}{0.357} = 2.8011$$

Now $P(z < -2.8011) = P(z > 2.8011)$

$$= 1 - P(z < 2.8011)$$

$$= 1 - 0.99744 = 0.00256$$

$P(z > 2.8011) = 0.00256$

Hence Probability coin is rejected $= 0.00256 + 0.00256 = 0.00512$.

Proportion rejected $= 0.5\%$

(c)   (i)   The population of weights of coins is normally distributed.

Looking at the sample data, the mean and median are very similar, and this suggests a symmetrical distribution. So we can say that the sample also follows a normal distribution.

(ii) The null hypothesis is that the mean weight of all the coins in the sample is equal to 7.12 g.

The alternative hypothesis is that the mean weight of all the coins in the sample is less than 7.12 g

Hence we can write, $H_0 : \mu = 7.12$

$$H_1 : \mu < 7.12$$

Need to calculate the *p*-value

$$p\text{-value} = P(\bar{x} \le 6.89) \text{ given the null hypothesis}$$

$$= P\left(z \le \frac{6.89 - 7.12}{\frac{0.357}{\sqrt{30}}}\right)$$

$$= P(z \le -3.52874)$$

$$= 0.00022$$

Now as *p*-value < 0.01, the null hypothesis H_0 is rejected.

There is strong evidence to suggest that the mean weight of a batch of 2p coins is less than 7.12 g.

This means the probability that the mean is less than 6.89 given the null hypothesis. Notice the way the mean value needs to be converted to a *z*-value using the formula

$$z = \frac{\bar{X} - \mu}{\frac{\sigma}{\sqrt{n}}}$$

BOOST
Grade ⇧⇧⇧⇧

It is not enough to say that the null hypothesis is rejected. You must make a comment about the mean in this case and make it relevant to the context of the question.

Active Learning

For this active learning you have to find out about the properties of normal distributions using the Internet.

Here is what you have to find out about and produce.

(a) Find out about examples of distributions that are normally distributed. Here are a couple to start you off.

- The weights of students in a sixth form.
- The lifespan in years of a particular breed of dog.

You should aim for at least 10 examples.

(b) If you were given some summary statistics that typically include the following list, explain the things you would be looking for in these statistics in order to decide whether the data was normally distributed or not.

- Summary statistic list
- Minimum value
- Lower quartile
- Mean
- Median
- Upper quartile
- Maximum value
- Standard deviation

(c) If data that was normally distributed was used to create a histogram, draw a diagram showing the shape of the histogram.

Test yourself

1. A supermarket manager thinks that the number of shoppers in the store each week is related to the number of special offers in the store. She takes samples over a period of 15 weeks and finds that the product moment correlation coefficient is 0.55. Carry out a hypothesis test at the 5% level of significance to decide whether the number of shoppers per week is correlated to the number of special offers.   [6]

2. A random sample of 36 observations were taken from a normal distribution with mean 60 and standard deviation 3 in order to test the hypotheses $H_0 : \mu = 60$ against $H_1 : \mu > 60$. Carry out this hypothesis test if the sample mean was found to be 61.   [5]

3. The mean lifespan for a particular breed of dog is 10 years with a standard deviation of 2 years. The lifespan has been found to be normally distributed. It is thought that the lifespan is greater than 10 years so a sample of 25 dogs of this breed was used to find the mean lifespan and it was found to be 11 years.
    Determine the p-value and interpret your result in context.   [4]

4. A machine fills sacks with fertiliser with the weight of fertiliser being normally distributed with a mean of μ kg and a standard deviation of 1.8 kg. If $\mu \neq 12$, the machine needs adjusting.
    A sample of 20 sacks was taken, and the mean weight of fertiliser was found to be 11.8 kg with a standard deviation of 1.8 kg. Determine whether or not the machine needs adjusting.   [5]

5. LED light bulbs have lifetimes that are normally distributed with a mean of 15 000 hours and standard deviation 2500 hours. A large batch of bulbs was made.
    A sample of 100 bulbs was taken and the mean lifetime of the bulbs found.
    (a) It has been decided that if the mean lifetime of the sample was 14 500 hours or less, then the whole batch would be considered substandard and discarded. Determine the significance level used for a test to decide whether the batch was substandard.   [3]
    (b) The significance level is to be set at 1%. Find the critical value for the lifetime of the bulbs.   [3]

6. The weights of suitcases to be loaded onto an aircraft are known to be normally distributed with a mean of 20.5 kg and a variance of 0.8 kg.
    A sample of 10 suitcases was taken and their weights measured and the mean weight was found to be 19.6 kg.
    The management of the airline suggest that the mean weight of suitcases has decreased.
    Assuming the variance is still the same, test the management's claim at the 1% significance level.   [5]

7   One metre lengths of fishing line have breaking strains that are normally distributed with a mean of 8 kg and a standard deviation of 0.9 kg. A new manufacturing process is to be used which also produces lines whose breaking strains are normally distributed but now with an increased mean but the same standard deviation of 0.9 kg. 30 one metre lengths were taken after being produced by the new process. The mean breaking strain of the sample was found to be 8.2 kg.

The production manager wants to find out whether the new process does, in fact, increase the breaking strain of the line.

(a) State suitable hypotheses for his investigation. [1]

(b) Determine the *p*-value of these results stating your conclusion in context. [8]

8   Amy wants to test if there is any correlation between the number of hours of sunshine and the mean temperature for her local town. Amy collects 40 pairs of readings and uses this data to calculate the product moment correlation coefficient which she found was 0.75.

(a) Write down suitable hypotheses she could use to test to see if there is any correlation. [2]

(b) Test at the 5% level of significance whether there is correlation between the number of hours of sunshine and mean temperature. [5]

Summary

Check you know the following facts:

Correlation coefficients

Have the letters ρ if based on a population or r if based on a sample.

They have values between and including -1 and 1.

> ρ or $r = 0$ means no correlation
>
> ρ or $r = 1$ means perfect positive correlation
>
> ρ or $r = -1$ means perfect negative correlation

Performing hypothesis testing using a correlation coefficient as a test statistic

Testing for positive correlation use $\mathbf{H}_0 : \rho = 0,\ \ \mathbf{H}_1 : \rho > 0$

Testing for negative correlation use $\mathbf{H}_0 : \rho = 0,\ \ \mathbf{H}_1 : \rho < 0$

Testing for any correlation (i.e. positive or negative) use $\mathbf{H}_0 : \rho = 0,\ \ \mathbf{H}_1 : \rho \neq 0$

Hypothesis testing for the mean of a normal distribution with a known, given or assumed variance

If X is normally distributed then $X \sim \mathrm{N}(\mu, \sigma^2)$ and the sample mean, $\overline{X}$, is normally distributed, so $\overline{X} \sim \mathrm{N}\!\left(\mu, \dfrac{\sigma^2}{n}\right)$

> This formula is used to work out z-values that apply to the standard normal distribution.

If $Z \sim \mathrm{N}(0, 1)$, then
$$Z = \frac{\overline{X} - \mu}{\frac{\sigma}{\sqrt{n}}}$$

Hence the z-value (i.e. the test statistic) will be:
$$z = \frac{\overline{x} - \mu}{\frac{\sigma}{\sqrt{n}}}$$

There are two methods that can be used for hypothesis testing:

- Using the z-value to find the probability (i.e. the p-value) and then compare this probability to the significance level.

- Use the significance level of the test to find the critical value and then see if the test statistic (i.e. the sample mean) lies inside or outside the critical region.

4 Kinematics for motion with variable acceleration

Introduction

You were introduced to kinematics in Topic 7 of the AS Applied course where you used graphs and the *suvat* equations to solve problems where the acceleration was constant.

In this topic, you will be considering motion in a straight line (called rectilinear motion) under variable acceleration. When the acceleration is variable, the velocity–time graph will be a curve and the gradient of the curve at a particular time, will be the acceleration at that time.

As the gradient of the curve varies with time, calculus is used to find the gradient and hence the acceleration at a particular time.

In a similar way, the gradient of a displacement–time graph is the velocity at a particular time and calculus can be used to find the gradient.

The area under a velocity–time graph represents displacement and integration can be used to find this area and hence the displacement.

This topic covers the following:

4.1 Using calculus for motion in a straight line in one dimension when the acceleration is not constant

4.1 Using calculus for motion in a straight line in one dimension when the acceleration is not constant

The graph of displacement against time shown below is not linear so the graph does not represent constant velocity. The gradient of a displacement–time graph represents the velocity so the velocity at point P is lower than that at point Q. The gradient and hence the velocity will depend on the time chosen.

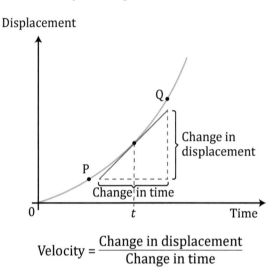

$$\text{Velocity} = \frac{\text{Change in displacement}}{\text{Change in time}}$$

Finding the velocity (v) from the displacement (s)

> To find the velocity from the displacement, you differentiate the expression for the displacement with respect to time.

The velocity at any time t is given by the gradient of the displacement–time graph.

Hence we have the important result $v = \dfrac{ds}{dt}$

To find the velocity, the time at which the velocity is found is substituted into the expression.

Example

1 A particle moves in a straight line in such a way that its displacement, s m at time t s is given by

$$s = 5t^3 - 4t^2 + 5$$

Find the velocity at $t = 1$ s

> Remember that to differentiate you multiply by the index and then reduce the index by one.

Answer

1    $s = 5t^3 - 4t^2 + 5$

$\qquad v = \dfrac{ds}{dt} = 15t^2 - 8t$

$\qquad\qquad = 15(1)^2 - 8(1)$

$\qquad\qquad = 7\ \text{m s}^{-1}$

Finding the acceleration (a) from the velocity (v)

> To find the acceleration from the velocity, you differentiate the expression for the velocity with respect to time.

The acceleration at any time t is given by the gradient of the velocity–time graph.

Hence, we have the important result $a = \dfrac{dv}{dt}$

Example

1 The velocity of a particle, v m s^{-1}, at time t s is given by the following

$$v = 3t^2 - 4t + 5$$

Find the acceleration at $t = 2$ s.

Answer

1 $v = 3t^2 - 4t + 5$

$a = \dfrac{dv}{dt}$

$\quad = 6t - 4$

When $t = 2$ s, $v = 6 \times 2 - 4 = 8$ m s^{-1}

Differentiating to find v and a

The following diagram summarises these results. Notice that if you are given an expression for the displacement and you are asked to find the acceleration, then you will need to differentiate the displacement expression twice.

Displacement (s) —Differentiate→ Velocity (v) —Differentiate→ Acceleration (a)

$\dfrac{ds}{dt}$ $\dfrac{dv}{dt}$

Example

1 A particle moves along a straight line in such a way that the displacement at time t is given by

$$s = 12t^3 - 6t^2 + 1$$

(a) Find an expression for the velocity, v, at time t.

(b) The particle is at rest twice. Find these two times.

(c) Find an expression for the acceleration, a, at time t.

Answer

1 (a) $v = \dfrac{ds}{dt}$

$\quad = 36t^2 - 12t$

Differentiating the expression for s with respect to t gives the velocity.

(b) When at rest, $v = 0$.

$0 = 36t^2 - 12t$

$0 = 12t(3t - 1)$

Solving gives $t = 0$ or $\dfrac{1}{3}$

(c) $a = \dfrac{dv}{dt}$

$\quad = 72t - 12$

Differentiating the expression for v with respect to t gives the acceleration, a.

BOOST

Grade ⇧⇧⇧⇧

When confronted by equations of this type, it is a common mistake to cancel the common factor t and then solve the resulting equation. If you do this you will lose one of the solutions (i.e. $t = 0$).

Remember that when integrating without limits, you must include the constant of integration.

We can now find the velocity at any time by substituting values into this equation for t.

Working backwards to find the velocity and the displacement

If you need to work backwards through the diagram shown in the previous section then you will need to reverse the effect of differentiating by integrating as the diagram below shows.

Displacement (s) ⟵ Integrate ⟵ Velocity (v) ⟵ Integrate ⟵ Acceleration (a)

$$s = \int v\,dt \qquad v = \int a\,dt$$

Remember that when you integrate without limits, you must include a constant of integration c. Usually there are some numerical values included in the question that can be substituted into the equation so the value of c can be found.

Finding the velocity from the acceleration

If you need to find the velocity from the acceleration, you integrate with respect to t.

Hence we have the important result $\quad v = \int a\,dt$

For example, if $a = 2 - t$ then $v = \int a\,dt = \int (2 - t)\,dt = 2t - \dfrac{t^2}{2} + c.$

Notice the inclusion of the constant of integration, c. It is necessary to find the value of c and this is done by substituting known values for v and t into the expression for v.

For example, if it is known that the object starts from rest, we can say that when $t = 0$, $v = 0$ so

$$0 = 2(0) - \frac{(0)^2}{2} + c \quad \text{and solving gives } c = 0.$$

The constant of integration, c, is now substituted into the expression to give

$$v = 2t - \frac{t^2}{2}$$

Finding the displacement from the velocity

To find the displacement from the velocity we integrate the velocity expression with respect to t.

Hence we have the important result $\quad s = \int v\,dt$

Carrying on with our example, to find the displacement we need to integrate the velocity expression, $v = 2t - \dfrac{t^2}{2}$ with respect to t.

Hence we have $\quad s = \int v\,dt = \int \left(2t - \dfrac{t^2}{2}\right)dt = \dfrac{2t^2}{2} - \dfrac{t^3}{3 \times 2} + c = t^2 - \dfrac{t^3}{6} + c$

Again we need to substitute known values of s and t into the expression to find the value of the constant c.

If we knew that when $t = 0$, $s = 0$, we have $0 = 0^2 - \dfrac{0^3}{6} + c$, giving $c = 0$.

The constant of integration, c, is now substituted into the expression to give

$$s = t^2 - \frac{t^3}{6}$$

4.1 Using calculus for motion in a straight line in one dimension when the acceleration is not constant

The following examples show these techniques.

Examples

1 A particle starts from rest and travels from O along the positive x-axis in a straight line. The particle has acceleration, a m s^{-2}, given by

$$a = 4t - 3t^2.$$

   (a) Find an expression for the velocity, v m s^{-1}, in time t.

   (b) Find an expression for the displacement, s m, in time t.

. . . .

Answer

1   (a) $v = \int a \, dt$

$$= \int (4t - 3t^2) \, dt$$

$$= \frac{4t^2}{2} - \frac{3t^3}{3} + c$$

$$= 2t^2 - t^3 + c$$

As the particle starts from rest, when $t = 0$, $v = 0$.

Hence we have $0 = 2(0)^2 - (0)^3 + c$

Solving gives $c = 0$.

Hence the expression for the velocity is

$$v = 2t^2 - t^3$$

> We need to find the value of the constant of integration, c, by substituting a pair of known values for v and t into this equation.

   (b) $s = \int v \, dt$

$$= \int (2t^2 - t^3) \, dt$$

$$= \frac{2t^3}{3} - \frac{t^4}{4} + c$$

When $t = 0$, $s = 0$.

$$0 = \frac{2(0)^3}{3} - \frac{(0)^4}{4} + c \text{ giving } c = 0.$$

Hence the expression for the displacement is

$$s = \frac{2t^3}{3} - \frac{t^4}{4}$$

---

2 An object accelerates from rest and at a time t seconds, its acceleration is given by

$$a = 5 - 0.1t \text{ m s}^{-2}.$$

   (a) Find the velocity after 10 seconds.

   (b) Show that the object travels at constant speed at $t = 50$ s.

   (c) Find the distance travelled in the first 50 seconds.

. . . .

Answer

2   (a) $v = \int a \, dt$

$$= \int (5 - 0.1t) dt$$

$$= 5t - \frac{0.1t^2}{2} + c$$

As the object starts from rest, we know when $t = 0$, $v = 0$.

So
$$0 = 5(0) - \frac{0.1(0)^2}{2} + c$$

Solving gives $c = 0$.

Hence we have $v = 5t - \frac{0.1t^2}{2}$

When $t = 10$, $v = 5 \times 10 - \frac{0.1 \times 10^2}{2}$

$$= 45 \text{ m s}^{-1}$$

(b) If the object travels at constant speed the acceleration would be zero.

When $t = 50$, $a = 5 - 0.1t$, so $a = 5 - 0.1(50) = 0$.

Hence object travels at constant speed when $t = 50$ s.

(c) $s = \int v \, dt$

$$= \int \left(5t - \frac{0.1t^2}{2} \right) dt$$

$$= \frac{5t^2}{2} - \frac{0.1t^3}{6} + c$$

As the object starts from rest when $t = 0$, $s = 0$.

Hence $c = 0$.

So we have $s = \frac{5t^2}{2} - \frac{0.1t^3}{6}$

When $t = 50$, $s = \frac{5(50)^2}{2} - \frac{0.1(50)^3}{6}$

$$= 6250 - 2083.3$$

$$= 4166.7 \text{ m}$$

---

3   A particle moves in a straight line with velocity v m s^{-1} at time t s, where

$$v = 4 \cos 2t$$

Calculate the exact distance travelled by the particle between $t = 0$ and $t = \frac{\pi}{6}$ s.

· ·

Answer

3 $s = \displaystyle\int_0^{\frac{\pi}{6}} v \, dt$

$$= \int_0^{\frac{\pi}{6}} (4 \cos 2t) dt$$

$$= \left[2 \sin 2t \right]_0^{\frac{\pi}{6}}$$

$$= \left[\left(2 \sin \frac{\pi}{3} \right) - (2 \sin 0) \right]$$

> Note the way the two times are used as the limits for the integration. Note that you are actually finding the area under the velocity–time graph between the two limits, which is equal to the distance travelled.

> Note that the integral of $\cos 2t$ is $\frac{1}{2} \sin 2t$.

$$= 2 \sin \frac{\pi}{3}$$

$$= 2 \times \frac{\sqrt{3}}{2}$$

$$= \sqrt{3} \text{ m}$$

BOOST

Grade ⇧⇧⇧⇧

Notice the word exact in the question. If you work out $\sqrt{3}$ as a decimal you will be giving an approximate answer. So leave the answer as $\sqrt{3}$.

Newton's second law of motion

When an unbalanced force, F, (called a resultant force) acts on a body, it produces an acceleration, a according to the following formula:

Force = mass × acceleration or $F = ma$.

Where F = force in N, a = acceleration in m s^{-2}, m = mass in kg

This equation may need to be used in questions on this topic and the following step by step illustrates a typical question.

Active Learning

Produce a poster containing the equations of motion, Newton's 2nd law of motion and the two diagrams on page 92 and take a picture of your finished product on your phone. You can keep referring to it when answering questions on kinematics with either constant or variable acceleration.

Step by STEP

A particle of mass 4 kg moves along the x-axis, starting, when $t = 0$, from the point where $x = 3$. At time t s, its velocity v m s^{-1} is given by

$$v = 12t^2 - 7kt + 1,$$

where k is constant.

When $t = 2$, the displacement of the particle from the origin is 16 m.

Calculate the magnitude of the force acting on the particle when $t = 5$.

Steps to take 👣👣👣

1   The information given in the question concerns the displacement. We need to therefore integrate the expression for v to find an expression for x.

2   Notice that two pairs of values for x and t are given. This is so that the value of k and the constant of integration can be found, otherwise the equation connecting x and t would have four unknowns.

3   When the values of x and t are substituted into the equations you will obtain a pair of equations in k and c which can be solved.

4   The value of k is substituted back into the velocity equation.

5   To find the acceleration the expression for the velocity is differentiated.

6   $t = 5$ is substituted into the expression for a to obtain a value for the acceleration.

7   $F = ma$ is used to obtain the magnitude of the force in Newtons.

$\cdots\cdots\cdots\cdots\cdots\cdots\cdots\cdots\cdots\cdots\cdots\cdots\cdots\cdots\cdots\cdots\cdots\cdots\cdots$

Answer

$$x = \int v\,dt$$

$$= \int (12t^2 - 7kt + 1)\,dt$$

$$= 4t^3 - \frac{7kt^2}{2} + t + c$$

Now, when $t = 0$, $x = 3$, hence $\qquad 3 = 4(0)^3 - \frac{7k(0)^2}{2} + (0) + c$ giving $c = 3$

Hence we can write $\qquad x = 4t^3 - \frac{7kt^2}{2} + t + 3$

Now, when $t = 2$, $x = 16$, hence $\qquad 16 = 4(2)^3 - \frac{7k(2)^2}{2} + 2 + 3$

$$16 = 32 - 14k + 5$$

> The value $k = \frac{3}{2}$ is substituted back into the velocity equation.

Solving gives $\qquad k = \frac{3}{2}$

Hence $\qquad v = 12t^2 - \frac{21}{2}t + 1$

$$a = \frac{dv}{dt} = 24t - 10.5$$

When $t = 5$, $\qquad a = 24(5) - 10.5 = 109.5$

Now $\qquad F = ma = 4 \times 109.5 = 438\text{ N}$

Example

1 A particle P, of mass 0.8 kg, moves along the x-axis so that its velocity at time t seconds is $v\text{ m s}^{-1}$, where $v = 4t^3 - 6t + 7$. Given that the displacement of P is 5 m from the origin when $t = 0$, find:

(a) The displacement of P from the origin when $t = 2$,

(b) The force acting on P when $t = 3$.

$\cdots\cdots\cdots\cdots\cdots\cdots\cdots\cdots\cdots\cdots\cdots\cdots\cdots\cdots\cdots\cdots\cdots\cdots\cdots$

Answer

1   (a) $x = \int v\,dt$

$$= \int (4t^3 - 6t + 7)\,dt$$

$$= 4\frac{t^4}{4} - \frac{6t^2}{2} + 7t + c$$

$$= t^4 - 3t^2 + 7t + c$$

Now, when $t = 0$, $x = 5$, hence $\quad 5 = (0)^4 - 3(0)^2 + 7(0) + c$, giving $c = 5$

Hence $\qquad x = t^4 - 3t^2 + 7t + 5$

When $t = 2$, $\qquad x = 2^4 - 3(2)^2 + 7(2) + 5 = 23\text{ m}$

(b) $a = \frac{dv}{dt}$

$$= 12t^2 - 6$$

When $t = 3$, $\quad a = 12t^2 - 6 = 12(3)^2 - 6 = 102\text{ m s}^{-2}$

> You need to remember $F = ma$ as it is not included in the formula booklet.

Force = mass × acceleration = $0.8 \times 102 = 81.6\text{ N}$

Test yourself

1. A particle moves along a straight line. At time t seconds, the displacement, s metres, of the particle from the origin is given by
$$s = 12t^3 + 9$$
   (a) Find an expression for the velocity in $m\,s^{-1}$ of the particle at time $t\,s$. [1]
   (b) Find the acceleration at time $t = 2$ seconds. [2]

2. A lorry accelerates from rest and at time t seconds, its acceleration $a\,m\,s^{-2}$ is given by
$$a = 3 - 0.1t \quad \text{until } t = 30\,s.$$
   (a) Find an expression in terms of t for the velocity of the lorry in $m\,s^{-1}$. [2]
   (b) Find the velocity of the lorry after 10 seconds in $m\,s^{-1}$. [1]
   (c) Explain what will happen to the lorry at $t = 30\,s$. [2]
   (d) Find the distance travelled in the first 30 seconds. [3]

3. A particle moves in a straight line and its velocity is $v\,m\,s^{-1}$, t seconds after passing the origin O where v is given by
$$v = 6t^2 + 4.$$
   Find the distance travelled between the times $t = 2\,s$ and $t = 5\,s$. [4]

4. A particle moves in a straight line and at time t seconds, it has velocity $v\,m\,s^{-1}$, where
$$v = 6t^2 - 2t + 8$$
   (a) (i) Find an expression for the acceleration of the particle at time t. [2]
       (ii) Find the acceleration of the particle when $t = 1\,s$. [2]
   (b) When $t = 0\,s$, the particle is at the origin. Find an expression for the displacement of the particle from the origin at time t. [3]

5. A particle moves along a straight horizontal line. Its velocity $v\,m\,s^{-1}$ at time $t\,s$, is given by
$$v = 2t(t - 6).$$
   (a) Find the set of values of t for which the velocity of the particle is negative. [3]
   (b) Find the total distance travelled by the particle in the interval $0 \le t \le 9$. [5]

Summary

Check you know the following facts:

Using calculus when the acceleration is not constant

Calculus has to be used to find s, v, or a when the acceleration is not constant and the following diagrams shows the processes involved.

Using differentiation

Displacement (s) →Differentiate $\dfrac{ds}{dt}$→ Velocity (v) →Differentiate $\dfrac{dv}{dt}$→ Acceleration (a)

Using integration

Displacement (s) ←Integrate $s = \int v\,dt$← Velocity (v) ←Integrate $v = \int a\,dt$← Acceleration (a)

Newton's 2nd law of motion

A resultant force, F N, produces acceleration, a m s^{-2}, on a mass, m kg, according to the formula

$$\text{Force} = \text{mass} \times \text{acceleration} \quad \text{or} \quad F = ma.$$

5 Kinematics for motion using vectors

Introduction

Topic 4 looked at kinematics with variable acceleration in one dimension. In this topic we will be looking at both motion with constant acceleration and then with variable acceleration for motion in a straight line in two and then three dimensions. You came across the use of vectors in the AS course and here you will be using vectors to describe motion in two or three dimensions using either modified *suvat* equations or calculus.

This topic covers the following:

5.1 Deriving the formulae for constant acceleration for motion in a straight line in two dimensions using vectors (i.e. adapting the *suvat* equations)

5.2 Equations of motion in two dimensions using vectors

5.3 Using calculus for motion in a straight line in two dimensions by making use of vectors

5.4 Using vectors in three dimensions

5.1 Deriving the formulae for constant acceleration for motion in a straight line in two dimensions using vectors (i.e. adapting the *suvat* equations)

If you have a particle moving in a straight line under constant acceleration, then you can use the adapted versions of the equations of motion with the unit vectors **i** and **j** to describe the motion.

For example, an initial velocity **u** in two dimensions could be written as $(3\mathbf{i} + 4\mathbf{j})$ m s^{-1} and this means there is a velocity of 3 m s^{-1} in the direction of the unit vector **i** and 4 m s^{-1} in the direction of the unit vector **j**. Note that the direction of the unit vector **i** is usually along the *x*-axis and the direction of the unit vector **j** is along the *y*-axis. You will be told in the question if this is not the case.

The equations of motion with constant acceleration in a straight line in two dimensions can be written as follows. Note the vector quantities are in bold and are not in italics.

$$\mathbf{v} = \mathbf{u} + \mathbf{a}t$$

$$\mathbf{s} = \mathbf{u}t + \frac{1}{2}\mathbf{a}t^2$$

$$\mathbf{v}^2 = \mathbf{u}^2 + 2\mathbf{a}\mathbf{s}$$

$$\mathbf{s} = \frac{1}{2}(\mathbf{u} + \mathbf{v})t$$

s = displacement vector
u = initial velocity vector
v = final velocity vector
a = acceleration vector
t = time scalar

Note that none of the equations of motion are included in the formula booklet so you will need remember them and also remember which of the letters represent vectors in each equation. Also, the vector quantities are emboldened and are not in italics. The only scalar quantity in these equations is time, *t*.

We also need to adapt Newton's 2nd law of motion to enable its use with vectors:

Force = mass × acceleration or $\mathbf{F} = m\mathbf{a}$

> Force and acceleration are both vectors but mass is a scalar.

5.2 Equations of motion in two dimensions using vectors

The formulae for the equations of motion adapted for vectors can be used in a similar way to the normal equations of motion except the variables that are vectors (**u**, **v**, **a** and **s**) are inserted in terms of the unit vectors **i** and **j** for problems in two dimensions.

Before using the equations of motion you must make sure that the acceleration is constant.

For example, $(3\mathbf{i} + 4\mathbf{j})$ m s^{-2} is a constant acceleration as its value does not change with time. The equations of motion can be used with accelerations like this. However, $(3t^2\mathbf{i} + 4t\mathbf{j})$ m s^{-2} is a variable acceleration as its value changes with time, as *t* appears in the expression. The equations of motion should not be used to solve problems with a variable acceleration. Instead we use calculus to solve these problems, as you saw in Topic 4.

Finding the scalar equivalent or magnitude of a vector quantity

Suppose you have a velocity, **v** given by **v** = 12**i** − 5**j**

The magnitude of this velocity is its speed and it can be calculated in the following way:

$$\text{Speed} = \sqrt{12^2 + (-5)^2} = \sqrt{169} = 13 \text{ m s}^{-1}$$

In general, if you have a vector and you are asked to find the magnitude (i.e. size) of the vector, you can use the following:

$$\text{Vector} = a\mathbf{i} + b\mathbf{j}$$

$$\text{Magnitude of vector} = \sqrt{a^2 + b^2}$$

The magnitude of a vector can be written with a modulus sign around the vector like this:

$|\mathbf{v}|$ for the magnitude of a velocity (i.e. the speed)

$|\mathbf{F}|$ for the magnitude of a force

$|\mathbf{a}|$ for the magnitude of an acceleration

Examples

1 A particle moves on a horizontal plane in a straight line such that its velocity is given by

$$\mathbf{v} = (12\mathbf{i} - 5\mathbf{j}) \text{ m s}^{-1}.$$

Find the speed of the particle.

· ·

Answer

1 Speed, $|\mathbf{v}| = \sqrt{12^2 + (-5)^2}$

$\qquad = \sqrt{169} = 13 \text{ m s}^{-1}$

---

2 A particle moves on a horizontal plane along a straight path. At time $t = 0$ s, the particle sets off from the origin O, and t seconds later it has a displacement of $(5\mathbf{i} - 2\mathbf{j})$ m from the origin. Find its distance from the origin at this time, giving your answer to two decimal places.

· ·

Answer

2 Distance, $|\mathbf{s}| = \sqrt{(5)^2 + (-2)^2}$

$\qquad = \sqrt{29}$

$\qquad = 5.39 \text{ m s}^{-1}$ (2 d.p.)

---

3 An object is accelerated with a constant acceleration of $(3\mathbf{i} + 4\mathbf{j})$ m s^{-2}.

Find the magnitude of this acceleration.

· ·

Answer

3 Magnitude of the acceleration, $|\mathbf{a}| = \sqrt{(3)^2 + (4)^2}$

$\qquad = \sqrt{25} = 5 \text{ m s}^{-2}$

4 An object of mass 4 kg is moving on a horizontal plane under the action of a constant force 4**i** – 12**j** N. At time $t = 0$ s, its position vector is 7**i** – 26**j** with respect to the origin O and its velocity vector is –**i** + 4**j**.

(a) Determine the velocity vector of the object at time $t = 5$ s.

(b) Calculate the distance of the object from the origin when $t = 2$ s.

> We know the force is constant because the force vector has no terms containing t in it.

Answer

4 (a) If the force is constant, then the acceleration will be constant.

Force = mass × acceleration (i.e. **F** = m**a**)

Hence $\mathbf{a} = \dfrac{\mathbf{F}}{m} = \dfrac{4\mathbf{i} - 12\mathbf{j}}{4} = \mathbf{i} - 3\mathbf{j}$

Using $\qquad\qquad \mathbf{v} = \mathbf{u} + \mathbf{a}t$

$\mathbf{v} = (-\mathbf{i} + 4\mathbf{j}) + (\mathbf{i} - 3\mathbf{j}) \times 5$

$= -\mathbf{i} + 4\mathbf{j} + 5\mathbf{i} - 15\mathbf{j}$

$= 4\mathbf{i} - 11\mathbf{j}$

(b) Using $\qquad\qquad \mathbf{s} = \mathbf{u}t + \dfrac{1}{2}\mathbf{a}t^2$

$\mathbf{s} = (-\mathbf{i} + 4\mathbf{j}) \times 2 + \dfrac{1}{2} \times (\mathbf{i} - 3\mathbf{j}) \times 4$

$= -2\mathbf{i} + 8\mathbf{j} + 2\mathbf{i} - 6\mathbf{j}$

$= 2\mathbf{j}$

Now at $t = 0$, the position vector is 7**i** – 26**j** so this vector needs to be added to the displacement vector travelled in 2 s to find the total displacement from the origin.

Hence displacement from the origin = 7**i** – 26**j** + 2**j** = 7**i** – 24**j**

Distance = $\sqrt{7^2 + (-24)^2}$ = 25 m

> **BOOST**
> **Grade** ⇧⇧⇧⇧
>
> Always check with the question to make sure that you are finding either the scalar or the vector. Here you are asked for distance which is the scalar quantity.

5.3 Using calculus for motion in a straight line in two dimensions by making use of vectors

In some problems, the acceleration varies, so the equations of motion cannot be used. To solve such problems, a method involving calculus is used.

Suppose the velocity of a particle is given by $\mathbf{v} = 2t\mathbf{i} - 4t^2\mathbf{j}$, you can see that this velocity contains t and as t varies, the velocity varies. If the velocity–time graph was drawn, it would not be linear, so the gradient and hence the acceleration would change with time. This means that calculus must be used to solve problems when the velocity changes with time.

If you had a problem where a particle was subjected to an acceleration given by $\mathbf{a} = (2t - 3)\mathbf{i} - (1 + 2t^2)\mathbf{j}$, you can see that the acceleration varies with time so again calculus must be used to solve problems concerning this particle.

We can adapt the formulae in the previous topic for use with vectors simply by ensuring the vectors are non-italicised and in bold like this:

5.3 Using calculus for motion in a straight line in two dimensions by making use of vectors

$$\mathbf{v} = \frac{d\mathbf{s}}{dt}$$

$$\mathbf{a} = \frac{d\mathbf{v}}{dt}$$

$$\mathbf{s} = \int \mathbf{v}\, dt$$

$$\mathbf{v} = \int \mathbf{a}\, dt$$

Examples

1 A particle moves on a horizontal plane such that its velocity vector $\mathbf{v}$ m s^{-1} at time t s is given by

$$\mathbf{v} = 7\sin 2t\,\mathbf{i} + 6\cos 3t\,\mathbf{j}.$$

(a) Find the acceleration vector of the particle at time t s.

(b) Given that when $t = 0$, the particle has position vector $(0.5\mathbf{i} + 3\mathbf{j})$ m, find the position vector of the particle when $t = \frac{\pi}{2}$.

· ·

Answer

1   (a) $\mathbf{v} = 7\sin 2t\,\mathbf{i} + 6\cos 3t\,\mathbf{j}$

$$\mathbf{a} = \frac{d\mathbf{v}}{dt} = 14\cos 2t\,\mathbf{i} - 18\sin 3t\,\mathbf{j}$$

> Note that the velocity–time graph for this equation would not be a straight line. So the gradient and hence the acceleration varies.

(b) $\mathbf{s} = \int \mathbf{v}\, dt = \int (7\sin 2t\,\mathbf{i} + 6\cos 3t\,\mathbf{j})\, dt$

$$= -3.5\cos 2t\,\mathbf{i} + 2\sin 3t\,\mathbf{j} + \mathbf{c}$$

When $t = 0$, $\mathbf{s} = 0.5\mathbf{i} + 3\mathbf{j}$

$$0.5\mathbf{i} + 3\mathbf{j} = -3.5\cos\big(2(0)\big)\mathbf{i} + 2\sin\big(3(0)\big)\mathbf{j} + \mathbf{c}$$

$$0.5\mathbf{i} + 3\mathbf{j} = -3.5\mathbf{i} + \mathbf{c}$$

> Note that the constant of integration is a vector.

Hence $\mathbf{c} = 4\mathbf{i} + 3\mathbf{j}$

When $t = \frac{\pi}{2}$, $\mathbf{s} = -3.5\cos\left(2\frac{\pi}{2}\right)\mathbf{i} + 2\sin\left(3\frac{\pi}{2}\right)\mathbf{j} + 4\mathbf{i} + 3\mathbf{j}$

> Note that the constants of integration in vectors are in vector form unless they are zero.

$$= (4 + 3.5)\mathbf{i} + (3 - 2)\mathbf{j}$$

$$= 7.5\mathbf{i} + \mathbf{j}\ \text{metres}$$

---

2 A particle P, of mass 2 kg, is moving so that at time t s its velocity $\mathbf{v}$ m s^{-1} is given by

$$\mathbf{v} = (13t - 3)\mathbf{i} + (2 + 3t^2)\mathbf{j}.$$

At time $t = 0$ s, the position vector of the particle is $(2\mathbf{i} + 7\mathbf{j})$ m.

(a) Find the position vector $\mathbf{r}$ of P at time t s.

(b) Determine the acceleration $\mathbf{a}$ of P at time t s.

(c) Calculate the magnitude of the acceleration when $t = 2$ s.

Answer

2   (a) $\mathbf{r} = \int \mathbf{v}\, dt$

$$= \int \left[(13t - 3)\mathbf{i} + (2 + 3t^2)\mathbf{j} \right] dt$$

$$= \left(\frac{13t^2}{2} - 3t \right)\mathbf{i} + (2t + t^3)\mathbf{j} + c$$

t = 0 is substituted into the equation for the position vector in order to find the constant of integration, *c*.

When *t* = 0, $\mathbf{r} = (2\mathbf{i} + 7\mathbf{j})$

Hence, $2\mathbf{i} + 7\mathbf{j} = 0\mathbf{i} + 0\mathbf{j} + c$

Therefore $c = 2\mathbf{i} + 7\mathbf{j}$

Hence position vector of P at time *t* s $= \left(\frac{13t^2}{2} - 3t \right)\mathbf{i} + (2t + t^3)\mathbf{j} + 2\mathbf{i} + 7\mathbf{j}$

$$= \left(\frac{13t^2}{2} - 3t + 2 \right)\mathbf{i} + (t^3 + 2t + 7)\mathbf{j}$$

Pythagoras' theorem is used to work out the magnitude of the acceleration.

(b) $\mathbf{a} = \dfrac{d\mathbf{v}}{dt} = 13\mathbf{i} + 6t\mathbf{j}$

(c)   When *t* = 2 s, $\mathbf{a} = 13\mathbf{i} + 12\mathbf{j}$

Note that |**a**| is often used as a shorthand for 'the magnitude of', in this case the acceleration.

Magnitude of acceleration $= |\mathbf{a}| = \sqrt{13^2 + 12^2}$

$$= 17.69 \text{ m s}^{-2} \quad (2 \text{ d.p.})$$

5.4 Using vectors in three dimensions

Forces, displacements, velocities and accelerations acting in three dimensions can be expressed in terms of the unit vectors **i**, **j** and **k**.

To find the magnitude of a vector quantity with the vector expressed in the form $a\mathbf{i} + b\mathbf{j} + c\mathbf{k}$ we use:

$$\text{Magnitude} = \sqrt{a^2 + b^2 + c^2}$$

Note that as the exact value of the speed is asked for we should not give the answer as a decimal.

For example, if a particle is travelling in a straight line with a velocity given by $\mathbf{v} = (-3\mathbf{i} + 4\mathbf{j} + 5\mathbf{k})$ m s^{-1} then the exact value of the speed can be found using

$$\text{Speed} = \sqrt{(-3)^2 + 4^2 + 5^2} = \sqrt{50} = 5\sqrt{2} \text{ m s}^{-1}$$

The equations of motion can be used with problems in three dimensions if the acceleration is constant. For problems involving variable acceleration we need to use calculus.

Example

1   At time *t* = 0 s, the position vector of an object A is **i** m and the position vector of another object B is 3**i** m. The constant velocity vector of A is $2\mathbf{i} + 5\mathbf{j} - 4\mathbf{k}$ m s^{-1} and the constant velocity vector of B is $\mathbf{i} + 3\mathbf{j} - 5\mathbf{k}$ m s^{-1}. Determine the value of *t* when A and B are closest together and find the least distance between A and B.

. . . .

Answer

1   The position vector after time t s is the sum of the position vector and the displacement vector.

As both objects are travelling at constant velocity we can use:

Displacement vector = $\mathbf{v}t$

Hence the position vector, $\mathbf{s}$, from the origin is

$\mathbf{s} = \mathbf{p} + \mathbf{v}t$ (where $\mathbf{p}$ is the initial position vector)

Position vector of A, $\mathbf{s}_A = \mathbf{i} + (2\mathbf{i} + 5\mathbf{j} - 4\mathbf{k})t = (1 + 2t)\mathbf{i} + 5t\mathbf{j} - 4t\mathbf{k}$

Position vector of B, $\mathbf{s}_B = 3\mathbf{i} + (\mathbf{i} + 3\mathbf{j} - 5\mathbf{k})t = (3 + t)\mathbf{i} + 3t\mathbf{j} - 5t\mathbf{k}$

Now the vector representing the displacement between these two position vectors is given by $\mathbf{s}_B - \mathbf{s}_A$

$$\mathbf{s}_B - \mathbf{s}_A = [(3 + t)\mathbf{i} + 3t\mathbf{j} - 5t\mathbf{k}] - [(1 + 2t)\mathbf{i} + 5t\mathbf{j} - 4t\mathbf{k}]$$

$$= (2 - t)\mathbf{i} - 2t\mathbf{j} - t\mathbf{k}$$

As this is a displacement, we need to change this into the distance, s. Hence,

$$s^2 = (2 - t)^2 + 4t^2 + t^2$$

$$= 6t^2 - 4t + 4$$

Now, we need to find the value of t for which s^2 is a minimum

$$\frac{d(s^2)}{dt} = 12t - 4$$

For a minimum value, $\dfrac{d(s^2)}{dt} = 0$

Hence $12t - 4 = 0$, giving $t = \dfrac{1}{3}$

$$s^2 = 6t^2 - 4t + 4$$

When $t = \dfrac{1}{3}$, $s^2 = 6\left(\dfrac{1}{3}\right)^2 - 4\left(\dfrac{1}{3}\right) + 4 = \dfrac{10}{3}$

Least distance $= \sqrt{\dfrac{10}{3}} = 1.83$ m

Step by STEP

An object of mass 7 kg is moving on a horizontal plane under the action of a constant force $7\mathbf{i} - 21\mathbf{j}$ N. At time $t = 0$, its position vector is $12\mathbf{i} + 2\mathbf{j}$ with respect to the origin and its velocity is $-\mathbf{i} + 4\mathbf{j}$.

(a) Determine the velocity vector of the object at time $t = 3$ s.

(b) Calculate the distance from the origin when $t = 3$ s.

Steps to take

1   The acceleration will be constant, since the force applied to the object is constant and force $\propto$ acceleration. You can find the acceleration, $\mathbf{a}$, using the following formula $\mathbf{a} = \dfrac{\mathbf{F}}{m}$ by substituting in $\mathbf{F}$ and m.

2   Use the equation of motion (adapted for vectors) $\mathbf{v} = \mathbf{u} + \mathbf{a}t$ to find the velocity after 3 s.

3   Now find the displacement vector using $\mathbf{s} = \mathbf{u}t + \frac{1}{2}\mathbf{a}t^2$. Now this is the displacement from the position when $t = 0$ s, which is not at the origin. We need to add this displacement vector to the position vector to find the total displacement from the origin.

4   Now find the distance by finding the magnitude of the displacement from the origin. This will then be the distance from the origin, O.

Answer

(a) $\mathbf{a} = \dfrac{\mathbf{F}}{m}$

$\qquad = \dfrac{1}{7}\left(7\mathbf{i} - 21\mathbf{j}\right)$

$\qquad = \mathbf{i} - 3\mathbf{j}$

$\mathbf{v} = \mathbf{u} + \mathbf{a}t$

$\qquad = (-\mathbf{i} + 4\mathbf{j}) + 3(\mathbf{i} - 3\mathbf{j})$

$\qquad = 2\mathbf{i} - 5\mathbf{j}$

> It is important to note that this will give the displacement from the starting position, which is not the origin.

(b) $\mathbf{s} = \mathbf{u}t + \frac{1}{2}\mathbf{a}t^2$

$\qquad = 3(-\mathbf{i} + 4\mathbf{j}) + \dfrac{1}{2}\left(\mathbf{i} - 3\mathbf{j}\right)(3^2)$

$\qquad = -3\mathbf{i} + 12\mathbf{j} + 4.5\mathbf{i} - 13.5\mathbf{j}$

$\qquad = 1.5\mathbf{i} - 1.5\mathbf{j}$

Displacement from origin O $= 12\mathbf{i} + 2\mathbf{j} + 1.5\mathbf{i} - 1.5\mathbf{j}$

$= 13.5\mathbf{i} + 0.5\mathbf{j}$

Distance from origin $= \sqrt{13.5^2 + 0.5^2}$

$= 13.5 \text{ m} \quad \text{(3 s.f.)}$

> You have now found the displacement of the object from the origin at $t = 3$ s; however, the question asks for the distance so we need to use Pythagoras' theorem to find the distance which is a scalar quantity.

Test yourself

1 A particle of mass 0.5 kg is moving under the action of a single force **F** N, where
$$\mathbf{F} = (4t - 3)\mathbf{i} + (3t^2 - 5t)\mathbf{j}.$$
The velocity of the particle at time t s is **v** m s^{-1}. When $t = 0$, $\mathbf{v} = 8\mathbf{i} - 7\mathbf{j}$.
Find an expression for **v** in terms of t. [5]

2 A particle moves in a straight line with velocity v m s^{-1} at time t s, where
$v = 4 \cos 2t$.
Calculate the distance travelled by the particle between $t = 0$ and $t = \frac{\pi}{6}$ s. [4]

3 A particle moves on a horizontal plane so that at time t seconds its position vector **r** metres relative to a fixed origin O is given by
$$\mathbf{r} = (t + 2t^2)\mathbf{i} + (1.5t^2 - 2t)\mathbf{j}.$$
Show that the acceleration of the particle is constant and find its magnitude. [4]

4 A particle of mass 2 kg moves under the action of a constant force **F** N, where **F** is given by
$$\mathbf{F} = -3\mathbf{i} + 4\mathbf{j} - 5\mathbf{k}.$$
(a) Find the magnitude of the acceleration of the particle. [3]
(b) Given that at time $t = 0$ seconds, the position vector of the particle is $2\mathbf{i} - 7\mathbf{j} + 9\mathbf{k}$ and it is moving with velocity $3\mathbf{i} - 2\mathbf{j} + \mathbf{k}$, find the position vector of the particle when $t = 2$ seconds. [3]

5 A particle of mass 3 kg moves in the xy plane in such a way that its position vector, **r**, at time t s is given by
$$\mathbf{r} = (4t^2 + 3)\mathbf{i} + (2 - 7t)\mathbf{j}.$$
(a) Show that the particle is moving under the action of a constant force. [3]
(b) Find the magnitude of this constant force. [2]

6 A particle moves on a horizontal plane such that its velocity vector **v** m s^{-1} at time t s is given by: $v = 2 \sin 3t\,\mathbf{i} + 4 \cos 3t\,\mathbf{j}$.
(a) Find the acceleration vector of the particle at time $t = \frac{\pi}{3}$ s. [3]
(b) Given that when $t = 0$ s, the particle has position vector $(\mathbf{i} + 2\mathbf{j})$ m, find the position vector of the particle when $t = \frac{\pi}{3}$ s. [4]

Summary

Check you know the following facts:

Formulae for constant acceleration for motion in a straight line using vectors

$$\mathbf{v} = \mathbf{u} + \mathbf{a}t$$

$$\mathbf{s} = \mathbf{u}t + \frac{1}{2}\mathbf{a}t^2$$

$$\mathbf{v}^2 = \mathbf{u}^2 + 2\mathbf{a}\mathbf{s}$$

$$\mathbf{s} = \frac{1}{2}(\mathbf{u} + \mathbf{v})t$$

$\mathbf{s}$ = displacement
$\mathbf{u}$ = initial velocity
$\mathbf{v}$ = final velocity
$\mathbf{a}$ = acceleration
t = time

Newton's 2nd law of motion

$$\text{Force} = \text{mass} \times \text{acceleration or } \mathbf{F} = m\mathbf{a}$$

Using calculus for motion in a straight line using vectors

For questions involving variable acceleration we make use of the following equations:

$$\mathbf{v} = \frac{d\mathbf{s}}{dt}$$

$$\mathbf{a} = \frac{d\mathbf{v}}{dt}$$

$$\mathbf{s} = \int \mathbf{v}\, dt$$

$$\mathbf{v} = \int \mathbf{a}\, dt$$

Finding the magnitude of a vector

- The magnitude of displacement is distance (a scalar).
- The magnitude of velocity is speed (a scalar).
- The magnitude of acceleration or force does not have its own name.
- For a vector in two dimensions such as vector, $\mathbf{v} = a\mathbf{i} + b\mathbf{j}$:
  magnitude of vector, $|\mathbf{v}| = \sqrt{a^2 + b^2}$
- For a vector in three dimensions such as vector, $\mathbf{v} = a\mathbf{i} + b\mathbf{j} + c\mathbf{j}$:
  magnitude of vector, $|\mathbf{v}| = \sqrt{a^2 + b^2 + c^2}$

6 Types of force, resolving forces and forces in equilibrium

Introduction

There are a number of different types of force you need to know about and these are weight, friction, normal reaction, tension and thrust.

A single force can be replaced by two different forces acting at right angles to each other and this process is called resolving a force. Resolving forces allows problems where several forces are acting to be solved. This topic also covers the composition of forces, which allows two or more forces to be replaced by a single equivalent force. Again this allows problems involving several forces to be solved.

Also included in this topic is the principle of moments, which allows problems involving systems of forces in equilibrium to be solved.

This topic covers the following:

6.1 Types of force

6.2 Modelling assumptions

6.3 Composition and resolution of forces

6.4 Equilibrium of a particle under the action of coplanar forces

6.5 The moment of a force about a point

6.6 Equilibrium of a rigid body under the action of parallel forces

6.1 Types of force

There are five different types of force you need to know about:

Weight – is the gravitational force between a particle of mass m kg and the Earth. It acts vertically down from the centre of mass of the particle to the centre of the Earth and its size is given by

$$\text{Weight} = mg$$

Where:

> m is the mass in kg and g is the acceleration due to gravity, which has the value 9.8 m s^{-2}.

The weight acts vertically down from the centre of mass of a body.

Friction – is a force that opposes motion and its size depends on how rough or smooth the surfaces are that are in contact. The force of friction can have any value up to a maximum value. Friction is dealt with in more detail in Topic 7.

Normal reaction – the force between a body and a surface when the two are in contact with each other. The normal reaction acts in a direction perpendicular to the surface. The normal reaction is the reaction force according to Newton's 3rd law of motion.

Tension – the resistive force in a string that resists the tendency for the string to extend.

Thrust – the resistive force provided by a spring or rod. Thrust acts in a direction so as to oppose the force either compressing or extending the spring or rod.

> The centre of mass is the point at which the entire mass of a body may be considered concentrated.

6.2 Modelling assumptions

Systems of forces are quite complicated in reality, and to make them easier to understand and describe mathematically by making a model we need to simplify things. This simplification involves making some assumptions and these are outlined here. You may be asked about modelling assumptions you have made when answering examination questions.

Strings are assumed to be:

- Light – which means they do not have a mass and hence a weight of their own.

- Inextensible – the string does not stretch, which means that the tension in the string remains constant.

Objects/masses are assumed to be:

- Particles – which means that they can be considered to be point masses whose dimensions are negligible.

Rods are assumed to be:

- Uniform along their length, unless you are told otherwise, so that the centre of mass of the rod is at the centre of the rod.

If we had to consider every factor/variable when we model arrangements of rods, strings and masses, things would become very complicated. We therefore need to make some modelling assumptions to simplify things.

Use the Internet to do some research into modelling assumptions in mechanics. After you have done this research, produce a list of the modelling assumptions you have found.

Active Learning

6.3 Composition and resolution of forces

Resolution of forces

A force F can be replaced by a pair of forces, called the components of F, with both of these components acting at right angles to each other as shown in the following diagram:

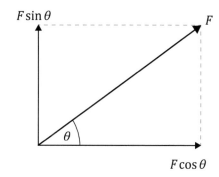

One component acts at an angle θ to the original force and has a size of $F \cos \theta$ and the other force acts at right angles to this component and has a size of $F \sin \theta$.

Sometimes you will be asked to find the resultant force, F, of two forces acting at right angles to each other and this is how to do it:

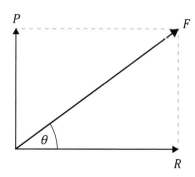

The size of the resultant force, F, is found by using Pythagoras' theorem, so we have

$$F^2 = P^2 + R^2$$

The direction of F can be referenced using the angle θ measured to the horizontal.

The angle θ can be found using

$$\tan \theta = \frac{P}{R} \qquad \text{so} \qquad \theta = \tan^{-1}\left(\frac{P}{R}\right)$$

Composition of forces

Composition of forces is finding a single force which is equivalent to two or more given forces acting in given directions.

Here is an example. You are asked to find the single force that would replace the three forces shown in the diagram.

> You can see from the diagram that the component of the 5 N force in the horizontal direction will add to the 6 N force. The vertical component of the 5 N force will be unable to balance the larger 8 N force acting down so the net force in this direction will be downwards.

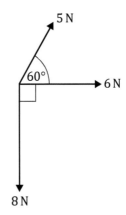

Component of the 5 N force in the horizontal direction = 5 cos 60° = 2.5 N

Resolving horizontally we obtain, net resultant force   = 6 + 2.5 = 8.5 N (to the right)

Component of the 5 N force in the vertical direction   = 5 sin 60° = 4.3301 N

Resolving vertically, we obtain, net resultant force   = 8 − 4.3301 = 3.6699 N
(downwards)

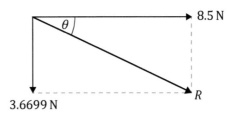

> Remember that the two forces are at right angles to each other so Pythagoras' theorem and trigonometry can be used to work out lengths of sides and angles.

Resultant force R is found using Pythagoras' theorem:

$$R^2 = 8.5^2 + 3.6699^2$$

giving $R = 9.2584$ N.

$$\theta = \tan^{-1}\left(\frac{3.6699}{8.5}\right) = 23.4° \text{ (to the horizontal)}$$

> You will not be given Pythagoras' theorem or the trigonometric ratios, so you will need to remember them.

Example

1   The diagram shows a sign attached to a point A. It is supported by two light rods AB and AC.

The rod AC is horizontal and the rod AB is inclined at an angle of α to the horizontal, where sin α = 0.6.

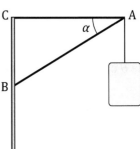

The mass of the sign is 12 kg. Calculate:

(a) The thrust in the rod AB.

(b) The tension in the rod AC.

Answer

1 (a)

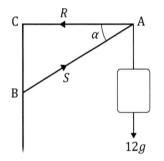 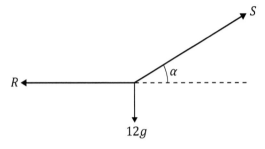

R is the tension in rod AC and *S* is the thrust in rod AB.

The forces acting at point A are in equilibrium.

Resolving vertically, we have

$$S \sin \alpha = 12g$$

$$S \times 0.6 = 12 \times 9.8$$

Thrust in AB, $S = 196 \text{ N}$

(b) Resolving horizontally, we have

$$S \cos \alpha = R$$

Tension in AC, $R = 156.8 \text{ N}$

Note that if, $\sin \alpha = 0.6$, $\cos \alpha = 0.8$

6.4 Equilibrium of a particle under the action of coplanar forces

A particle is said to be in equilibrium under the action of two or more forces, if the particle remains stationary. You saw previously that unbalanced forces produce a net or resultant force that causes a particle to be accelerated. When a particle is in equilibrium there is no resultant force and therefore no acceleration.

To solve a problem involving particles in equilibrium there are a number of steps to take:

- Draw an accurate diagram (if one is not already provided) and mark on all the forces acting.

- Resolve forces into two components which are perpendicular to each other.

- Equate to zero all the forces acting in one of the perpendicular directions.

- Equate to zero all the forces acting in the other perpendicular direction.

- Solve these equations to find the unknown forces or angles.

Examples

1   The particle in the diagram is in equilibrium under the action of three forces.

Resolve horizontally and vertically to find the size of forces P and R correct to two decimal places.

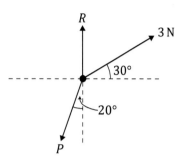

Here we have taken forces to the right and upward forces as being in the positive direction. Alternatively, you can just equate the upward and downward forces and also equate the forces to the right with the forces to the left.

Answer

1   Resolving horizontally, we obtain

$$3 \cos 30° - P \sin 20° = 0$$

Hence

$$P = \frac{3 \cos 30°}{\sin 20°} = 7.5963 \text{ N}$$

$$P = 7.60 \text{ N (correct to two decimal places)}$$

Resolving vertically, we obtain

$$R + 3 \sin 30° - P \cos 20° = 0$$

$$R + 1.5 - 7.5963 \cos 20° = 0$$

$$R = 5.6382 \text{ N}$$

$$R = 5.64 \text{ N (correct to two decimal places)}$$

Note that we have only been asked to find the size of the force and not its direction as well.

2   The diagram shows four horizontal forces acting at a point O. The forces are in equilibrium. Calculate the value of P and the size of the angle θ. Give each of your answers correct to one decimal place.

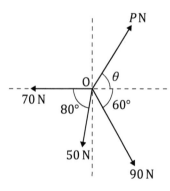

Answer

2   Resolving horizontally, and equating left and right forces, we obtain

$$P \cos \theta + 90 \cos 60° = 70 + 50 \cos 80°$$

Hence $$P \cos \theta = 33.6824$$ (1)

Do not round your intermediate answers off to one decimal place. Only round off your final answers.

Resolving vertically, we obtain

$$P \sin \theta = 90 \sin 60° + 50 \sin 80°$$

Hence $\qquad P \sin \theta = 127.1827 \qquad\qquad\qquad$ (2)

Dividing equation (2) by equation (1) gives

$$\tan \theta = \frac{127.1827}{33.6824} = 3.7759$$

Hence $\qquad\qquad \theta = 75.1666°$

So $\theta = 75.2°$ correct to one decimal place

> Remember that
> $$\frac{\sin \theta}{\cos \theta} = \tan \theta$$

Squaring equations (1) and (2) and then adding them together, we obtain

$$P^2 \sin^2 \theta + P^2 \cos^2 \theta = 127.1827^2 + 33.6824^2$$

$$P^2(\sin^2 \theta + \cos^2 \theta) = 127.1827^2 + 33.6824^2$$

$$P^2 = 127.1827^2 + 33.6824^2$$

Hence $\qquad\qquad P = 131.6\,\text{N}$ (correct to one decimal place)

> Here we use the result
> $$\sin^2 \theta + \cos^2 \theta = 1$$

Step by STEP

In the following question you are asked to find two tensions and this is an unstructured question as there is no guidance given as to the steps you need to take.

The diagram below shows an object of weight 110 N at a point C, supported by two cables and inclined at angles of 18° and 41° to the horizontal respectively.

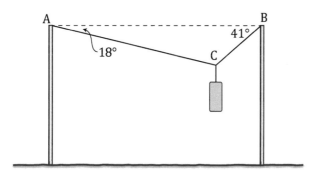

(a) Find the tension in AC and the tension in BC.

(b) State two modelling assumptions you have made in your solution.

Steps to take 👣

1   Look at the question and the diagram carefully. The system must be in equilibrium, so we can use resolution of forces.

2   The tensions in each string will be different, so we can call them T_{AC} and T_{BC}.

3   We can resolve the forces vertically and then horizontally to obtain two equations relating the tensions T_{AC} and T_{BC} which can then be solved simultaneously.

4   Look at the diagram carefully and look at the parts of the arrangement – the strings and the weight – and think of the modelling assumptions that could be applied to these.

Answer

Now follow the solution below.

(a)   Resolving horizontally, we obtain $T_{AC} \cos 18° = T_{BC} \cos 41°$

Resolving vertically, we obtain $T_{AC} \sin 18° + T_{BC} \sin 41° = 110$

Solving these equations simultaneously to find values for T_{AC} and T_{BC}. From the second equation:
$$T_{BC} = \frac{110 - T_{AC} \sin 18°}{\sin 41°}$$

Substituting this expression into the first expression in place of T_{BC} we obtain:

$$T_{AC} \cos 18° = \cos 41° \frac{110 - T_{AC} \sin 18°}{\sin 41°}$$

$$T_{AC} \cos 18° \sin 41° = 110 \cos 41° - T_{AC} \sin 18° \cos 41°$$

$$\cos 18° \sin 41° = \frac{110 \cos 41°}{T_{AC}} - \sin 18° \cos 41°$$

$$T_{AC} = \frac{110 \cos 41°}{\cos 18° \sin 41° + \sin 18° \cos 41°}$$

$$T_{AC} = 96.9 \, \text{N} \quad (3 \text{ s.f.})$$

> The value for T_{AC} is substituted into this equation.

$$T_{BC} = \frac{110 - T_{AC} \sin 18°}{\sin 41°}$$

$$T_{BC} = 122 \, \text{N} \quad (3 \text{ s.f.})$$

(b)   We have assumed that all of the forces on the object have been applied at one point, therefore the first assumption is that the object is a particle.

We haven't taken into consideration the weight of the cables, therefore the second assumption is that the cables are light.

6.5   The moment of a force about a point

> This is an anticlockwise moment as it produces an anticlockwise turning effect. The units of moments are N m.

A moment is the turning effect of a force. The moment of a force about a point P is the product of the magnitude of the force and the perpendicular distance of the line of action of the force from the point P.

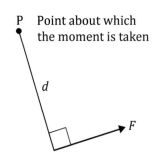

P   Point about which the moment is taken

d

F

Moment = Force × Distance

> Note that the distance must be the perpendicular distance from the point about which moments are taken, to the line of action of the force.

Moments can be either clockwise or anticlockwise depending on the direction of the turning effect they produce. We take one direction (usually clockwise) to be positive and the other direction to be negative. Moments are taken about a point to find the sizes of forces or their distances to a certain point by using the following principle of moments:

> When a body is in equilibrium, the sum of the clockwise moments is equal to the sum of the anticlockwise moments.

6.6 Equilibrium of a rigid body under the action of parallel forces

If a rigid body, such as a rod, under the action of parallel coplanar forces is in equilibrium then the following is true:

1. The resultant force in any direction is zero.

2. The sum of the moments about any point is zero.

Examples

1 The diagram shows a uniform rod AB of length 2 m and mass 10 kg. The rod is resting horizontally in equilibrium on two supports, one at point A and the other at C where AC = 1.5 m.

Calculate the reactions at A and C.

· ·

Answer

1

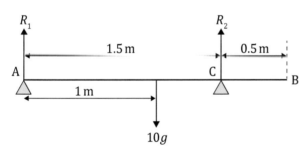

As the rod is in equilibrium, resolving vertically, we obtain

$$R_1 + R_2 = 10g \qquad (1)$$

Taking moments about point A, we obtain

$$10g \times 1 = R_2 \times 1.5$$

$$R_2 = 65.33 \text{ N}$$

Substituting this value of R_2 into equation (1) we obtain

$$R_1 + 65.33 = 10 \times 9.8$$

$$R_1 = 32.67 \text{ N}$$

Draw an accurate diagram marking all the forces acting along with the distances given in the question.

As the rod is uniform, the weight of the rod $10g$ may be supposed to act downwards at the centre of the rod.

The total of the upwards forces will equal the total of the downward forces.

Point A is chosen because R_1 acts through A and will have no moment about A (i.e. because its distance to A is zero). Using this point, we can find R_2 easily. Note that anticlockwise and clockwise moments are equal.

2   A uniform rod AB is suspended horizontally from the ceiling by means of two vertical light inextensible strings XA and YB of equal length.

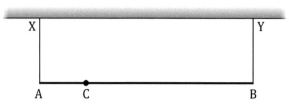

The rod AB has mass 6 kg and length 1.4 m. A particle, of mass 10 kg, is attached to the rod at point C, where AC = 0.3 m. Calculate the tension in each of the strings XA and YB.

> The tensions in the strings are not the same and they act vertically upwards on the rod.

Answer

2

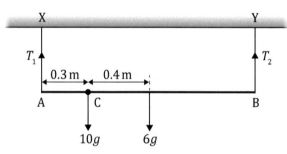

Let the tension in string   YB = T_2

Taking moments about point A, we have

$$T_2 \times 1.4 = (6 \times 9.8 \times 0.7) + (10 \times 9.8 \times 0.3)$$

> The choice of A is made because T_1 has no moment about A.

giving $T_2 = 50.4$ N

Let the tension in string   XA = T_1

Distance CB = 1.4 − 0.3 = 1.1

Taking moments about point B, we have

$$T_1 \times 1.4 = (6 \times 9.8 \times 0.7) + (10 \times 9.8 \times 1.1)$$

> You could alternatively have resolved vertically to obtain an equation containing both T_1 and T_2. The value of T_2 could be substituted into this equation to find T_1.

giving $T_1 = 106.4$ N

3   The diagram shows a body, of mass 65 kg, attached to the end B of a uniform rigid rod AB of length 4 m. The mass of the rod is 35 kg. The rod is held horizontally in equilibrium by two smooth cylindrical pegs, one at A and another at C, where AC = 1.2 m.

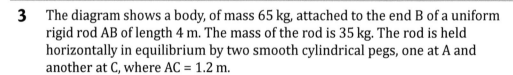

(a)   Write down the moment of the weight of the rod about the point A. State your units clearly.

(b) Find the forces exerted on the rod at A and C.

Answer

3 (a)

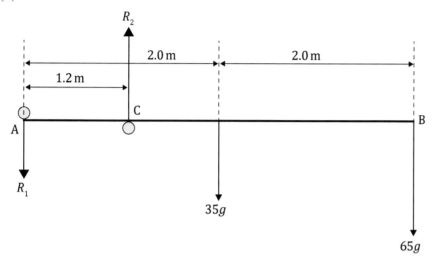

> To decide on the directions of the reactions R_1 and R_2 think about the forces exerted by the rod on the pegs A and C which will be upwards and downwards respectively. The forces on the rod are in an opposite direction according to Newton's 3rd law.

Moment of the weight about A = $F \times d$

$$= 35g \times 2$$

$$= 686 \, \text{N m (in a clockwise direction)}$$

(b) As the rod is in equilibrium, resolving vertically, we obtain

$$R_2 = R_1 + 35g + 65g$$

$$R_2 = R_1 + 100g \qquad (1)$$

Taking moments about A, we obtain

$$R_2 \times 1.2 = 686 + 65g \times 4$$

Hence $R_2 = 2695 \, \text{N}$

Substituting this value of R_2 into equation (1), we obtain

$$2695 = R_1 + 980$$

Hence $R_1 = 1715 \, \text{N}$

---

4 A uniform rod AB, of mass 3 kg, has length 2 m. A particle of mass 5 kg is attached to the end A, and a particle of mass 2 kg is attached to the end B. The diagram shows the rod resting horizontally in equilibrium on a smooth support at the point C, where AC = x m.

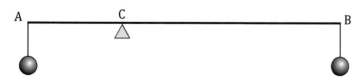

Calculate the magnitude of the reaction of the support at C and the value of x.

Answer

4

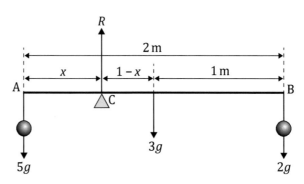

Resolving vertically we obtain

$$R = 5g + 3g + 2g$$

$$R = 10g$$

$$R = 98 \text{ N}$$

Taking moments about the pivot C

$$5gx = 3g(1 - x) + 2g(2 - x)$$

$$10x = 7$$

$$x = 0.7 \text{ m}$$

5　The diagram shows a uniform straight rod of length 4 m, resting horizontally in equilibrium on two supports at C and D. An object of mass 2.5 kg is suspended from point B.

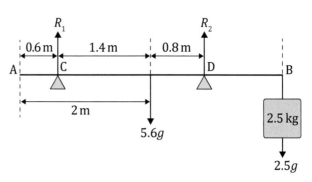

If the mass of the rod is 5.6 kg and distances AC = 0.6 m and AD = 2.8 m, calculate the magnitudes of the reactions at C and D.

Answer

5

Resolving vertically, we obtain

$$R_1 + R_2 = 5.6g + 2.5g$$

$$R_1 + R_2 = 79.38 \qquad\qquad (1)$$

Taking moments about point C, we obtain

$$(5.6g \times 1.4) + (2.5g \times 3.4) = R_2 \times 2.2$$

Solving gives $\qquad\qquad R_2 = 72.787\,\text{N}$

Substituting this value for R_2 into equation (1), we obtain

$$R_1 + 72.787 = 79.38$$

Hence $\qquad\qquad R_1 = 6.593\,\text{N}$

---

6 A light uniform rod AB has length 1.4 m. A particle of mass 5 kg is attached to end A, and a particle of mass 2 kg is attached to end B. The rod rests horizontally in equilibrium on a smooth support at C.

(a) Calculate the reaction of the support at C.

(b) Find the distance AC.

Answer

6 (a)

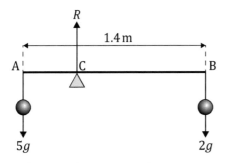

Resolving vertically, we obtain $\quad R = 5g + 2g$

$$= 7g$$

(b) Let the distance AC = x, so the distance CB = $1.4 - x$

Taking moments about C, we obtain

$$5gx = 2g(1.4 - x)$$

$$5x = 2.8 - 2x$$

$$7x = 2.8$$

$$x = 0.4\,\text{m}$$

Hence $\qquad\qquad AC = 0.4\,\text{m}$

Test yourself

1   Three horizontal forces of magnitudes 10 N, 8 N and 6 N act on a particle in the directions shown in the diagram.

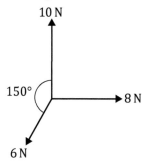

Find the magnitude of the resultant force and the angle between the resultant force and the 8 N force.   [4]

2   The diagram shows a uniform rod AB of length 2 m and mass 4 kg with a particle of mass 0.5 kg attached at A. The rod is resting horizontally in equilibrium on two smooth supports at points P and Q of the rod, where AP = 0.4 m and AQ = 1.4 m.

Calculate the reactions at P and Q.   [4]

3   A uniform rod AB of length 1.8 m and mass 5 kg is suspended horizontally from a rigid horizontal ceiling at points P and Q by two vertical, light, inextensible strings.

A particle of mass 2 kg is placed on the rod at point C which is 0.2 m from B. Calculate the tensions in each string.   [5]

4   Four horizontal forces of magnitude 6 N, 9 N, P N and Q N acting at a point are in equilibrium. Directions are as shown in the diagram.

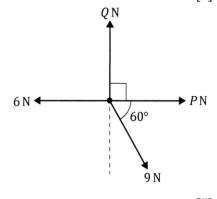

Find the value of P and the value of Q.   [5]

5 The diagram below shows an object of weight 160 N at a point C, supported by two cables AC and BC inclined at angles of 23° and 40° to the horizontal respectively.

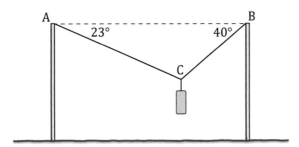

The tension in AC is T_1 N and the tension in BC is T_2 N.
(a) Show that $T_2 \approx 1.2T_1$. [2]
(b) Find the value of T_1. [3]
(c) State two modelling assumptions you have made in your solution. [2]

6 The diagram shows a plank AB, of mass 15 kg and length 2.8 m, being held in equilibrium with AB horizontal by means of two vertical ropes, one attached to the end A and the other attached to the end B. A man of mass 80 kg stands on the plank at point C, where AC = 0.9 m.

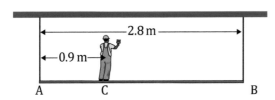

(a) Modelling the plank as a uniform rod, find the tensions in the ropes attached to the end A and the end B of the plank. [7]
(b) The plank is now modelled as a non-uniform rod. Given that the tension in the rope attached to A is 1.5 times the tension in the rope attached to B, determine the distance of the centre of mass of the plank from A. [5]

7

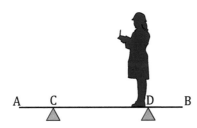

The diagram shows a uniform plank AB, of mass 20 kg and length 2.4 m, supported in horizontal equilibrium by two pivots, one at C and one at D. The distance AC and the distance DB are both 0.5 m. A person of mass 40 kg stands at a point which is 0.6 m from B.
(a) Calculate the magnitudes of the reaction at C and the reaction at D. [7]
(b) The person starts to walk towards A. Determine the greatest distance of the person from B if equilibrium is to be maintained. [3]

Summary

Check you know the following facts:

Resolution of forces

Replacing a single force with two forces acting at right angles

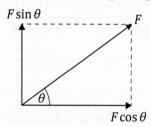

Force F can be replaced by two components at right angles to each other:

- A horizontal component $F \cos \theta$
- A vertical component $F \sin \theta$

Replacing two forces acting at right angles with a single force

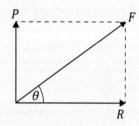

Using Pythagoras' theorem, so we obtain $F^2 = P^2 + R^2$

Hence $F = \sqrt{P^2 + R^2}$ and angle to the horizontal, $\theta = \tan^{-1}\left(\dfrac{P}{R}\right)$

The moment of a force about a point

$$\text{Moment} = \text{force} \times \text{distance}$$

The unit of moment is the newton metre (N m) and the distance must be the perpendicular distance from the point about which moments are taken to the line of action of the force.

Moments have a sense/direction (i.e. clockwise or anticlockwise).

The principle of moments:

When a body is in equilibrium, the sum of the clockwise moments is equal to the sum of the anticlockwise moments.

Equilibrium of a rigid body under the action of parallel coplanar forces

If a rigid body, such as a rod, under the action of parallel coplanar forces is in equilibrium then the following is true:

- The resultant force in any direction is zero.
- The sum of the moments about any point is zero.

7 Forces and Newton's laws

Introduction

In this topic you will be learning about Newton's laws of motion and how they can be applied to problems where several forces act. You will also come across frictional forces, which until now we have regarded as negligible as the surfaces touching have been smooth. You will look at the systems you looked at in AS except now the systems are more complex with the introduction of frictional forces and sometimes the particles being considered are moving or in equilibrium on a slope.

7.1 Newton's laws of motion

Newton's laws of motion were covered in the AS Applied course. Here is a reminder of the three laws.

Forces are vectors and, like all vectors, they have both magnitude (size) and direction. Forces are measured in newtons. When forces are unbalanced they are said to be not in equilibrium as there is a force called a resultant force acting. This resultant force produces an acceleration in the same direction as the resultant force.

Newton's 1st law – a particle will remain at rest or will continue to move with constant speed in a straight line unless acted upon by some external force. This means that if a particle is acted upon by unbalanced forces, then its velocity will change (i.e. it will accelerate or decelerate).

Newton's 2nd law – a resultant force produces an acceleration, according to the formula Force = mass × acceleration or $F = ma$

Newton's 3rd law – every action has an equal and opposite reaction. This means that if particle A exerts a force on particle B, then particle B will exert an equal and opposite force on particle A.

Resultant forces and their accelerations

When a number of coplanar forces act on a particle and the particle is not in equilibrium then there will be a resultant force and a resulting acceleration (or deceleration). To find the resultant force and its direction we find the component of all the forces in two directions that are perpendicular to each other. We can then use Pythagoras' theorem and trigonometry to find the magnitude and direction of the resultant force.

The resultant acceleration will always be in the same direction as the resultant force and can be calculated using Force = mass × acceleration.

Step by STEP

Four coplanar horizontal forces of magnitude 60 N, 85 N, 75 N and 40 N act on a particle P, of mass 5 kg, in the directions shown in the diagram, where $\tan \alpha = \frac{3}{4}$.

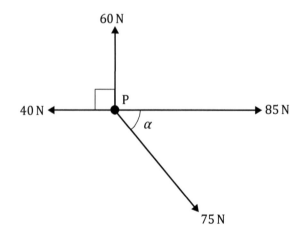

Calculate the magnitude of the acceleration of the particle P.

BOOST

Grade ⬆⬆⬆⬆

Make sure that you are clear about which direction you are taking as positive for the forces.

Steps to take

1 Work out the resultant force in the direction of the 85 N force and work out the resultant force in the vertical direction.

2 Use these two forces and work out the resultant of them using Pythagoras' theorem.

3 Use $F = ma$ to work out the acceleration which will be in the same direction as the resultant force.

. . . .

Answer

1 Resultant force in the direction of the 85 N force $= 85 + 75 \cos \alpha - 40$

$$= 45 + 75 \cos \alpha$$

Now $\tan \alpha = \frac{3}{4}$ so $\cos \alpha = \frac{4}{5}$

Resultant force in the direction of the 85 N force $= 45 + 75 \times \frac{4}{5}$

$$= 105\,\text{N (to the right)}$$

Resultant force in the direction of the 60 N force $= 60 - 75 \sin \alpha$

$$= 60 - 75 \times \frac{3}{5}$$

$$= 15\,\text{N (in the direction of the}$$
$$\text{original 60 N force)}$$

> Note that this is a 3:4:5 triangle or alternatively the length of the hypotenuse can be found using Pythagoras' theorem.

$R = \sqrt{105^2 + 15^2} = 106.066\,\text{N}$

Using $F = ma$ we have $106.066 = 5a$

$$\text{So} \quad a = 21.21\,\text{m s}^{-2}$$

> Note that only the magnitude of the acceleration is asked for, so there is no need to work out the angle θ marked on the diagram.

Examples

1 Particle P is subject to two coplanar forces of 10 N acting in the direction as shown in the diagram.

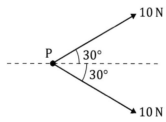

(a) Find the magnitude and direction of the resultant force.

(b) If the particle has a mass of 4 kg, find the magnitude of the acceleration of P.

1   (a)   Component force acting in the direction of the dotted line (taking to the right as the positive direction) = 10 cos 30° + 10 cos 30°

$$= 20 \cos 30°$$

$$= 17.32\,\text{N}$$

Component force acting in the direction at right angles to the dotted line   = 10 sin 30° − 10 sin 30° = 0

Resultant force = 17.32 N to the right in the direction of the dotted line.

(b) $F = ma$, so $a = \dfrac{F}{m} = \dfrac{17.32}{4} = 4.33\,\text{m s}^{-2}$

---

2   A particle P lies on a horizontal plane. Three horizontal forces of magnitude 7 N, 12 N and 16 N acting in directions as shown in the diagram, are applied to P.

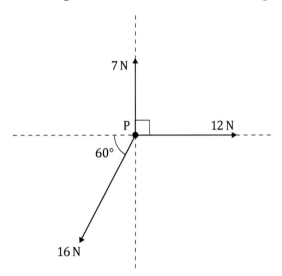

(a)   Show that the magnitude of the resultant of the three forces is approximately 7.9 N. Find the angle between the direction of the resultant and the direction of the 12 N force.

(b)   The particle P has mass 5 kg and the coefficient of friction between P and the plane is 0.1. Taking the magnitude of the resultant of the three forces to be 7.9 N, calculate the magnitude of the acceleration of P.

2   (a)   Component of the 16 N force in the opposite direction to the 12 N force

$$= 16 \cos 60° = 8\,\text{N (to the left)}$$

Net resultant force = 12 − 8 = 4 N
                                        (to the right, in the direction of the original 12 N force)

Component of the 16 N force acting in the opposite direction to the 7 N force

$$= 16 \sin 60° = 13.8564\,\text{N}$$

Net resultant vertical force = 13.8564 − 7 = 6.8564 N (downwards)

Resultant force F is found using Pythagoras' theorem

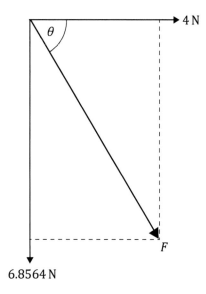

6.8564 N

$$F = 4^2 + 6.8564^2$$

giving $\quad F = 7.9379 \text{ N} = 7.9 \text{ N}$ (correct to one decimal place)

$$\theta = \tan^{-1}\left(\frac{6.8564}{4}\right) = 59.7° \text{ (to the direction of the 12 N force)}$$

(b)

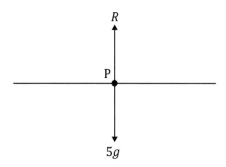

$5g$

The maximum frictional force depends on the normal reaction of the plane on the particle. Resolving vertically, we have

$$R = 5g$$

Solving gives $\quad R = 49 \text{ N}$

The driving horizontal force on the particle is the resultant found in part (a), i.e. 7.9 N, which is directly opposed by the maximum frictional force

$$\mu R = 0.1 \times 49 = 4.9 \text{ N}$$

Then, applying Newton's 2nd law to the particle in a horizontal direction, we have

$$5a = 7.9 - 4.9$$

giving $\quad a = 0.6 \text{ m s}^{-2}$

> R and $5g$ act in the vertical direction.

> The normal reaction is $5g$ (i.e. equal to the weight).
> All the other forces act in the horizontal plane.

3   An object of mass 80 kg is being dragged along a straight line AB by means of three horizontal forces of magnitude and direction as shown in the diagram. The resistance to the motion of the object is constant and of magnitude 16 N.

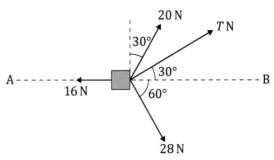

(a)   Show that $T = 8\sqrt{3}$.

(b)   Determine the magnitude of the acceleration of the object.

(c)   When the object is moving with a speed of 12 m s^{-1}, the three horizontal forces of 20 N, 28 N, and T N are removed. Calculate the time taken for the speed of the object to reduce to 4 m s^{-1}.

. . . .

Answer

> Work in surds because what you are asked to prove contains a surd.

3   (a)   Resolving at right angles to AB, we obtain:

$$20 \cos 30° + T \sin 30° = 28 \sin 60°$$
$$20 \times \frac{\sqrt{3}}{2} + T \times \frac{1}{2} = 28 \times \frac{\sqrt{3}}{2}$$
$$T = 8\sqrt{3}$$

(b)   Applying Newton's 2nd law of motion taking to the right as the positive direction, we obtain:

$$20 \sin 30° + T \cos 30° + 28 \cos 60° - 16 = 80a$$
$$20 \times \frac{1}{2} + 8\sqrt{3} \times \frac{\sqrt{3}}{2} + 28 \times \frac{1}{2} - 16 = 80a$$
$$a = 0.25 \text{ m s}^{-2}$$

(c)   Applying Newton's 2nd law of motion taking to the right as the positive direction, we obtain:

$$-16 = 80a$$
$$a = -0.2 \text{ m s}^{-2}$$
$$v = u + at$$
$$4 = 12 - 0.2t$$
$$t = 40 \text{ s}$$

BOOST

Grade ⇧⇧⇧⇧

Remember that the equations of motion are not included in the formula booklet.

Active Learning

Produce a brief guide to forces acting on a particle when:

(a)   The forces are in equilibrium and you are required to find the magnitude and direction of an unknown force.

(b)   There is an overall force acting on the particle and you want to find the magnitude and direction of the resultant.

Take a photo of your guide and use it for future revision.

7.2 Motion on an inclined plane

Before starting this section you should look back at the AS Applied book on the motion of particles connected by strings passing over pulleys or pegs which starts on page 157. In this topic things get slightly more complicated because one of the particles can be on an inclined plane and the plane may not be smooth so that frictional forces need to be considered.

When one particle is freely hanging and the other particle is on a smooth inclined plane

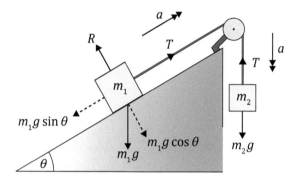

In this arrangement, the block on the inclined plane is accelerating up the slope. You can also get the situation where the mass would accelerate down the slope. For practice you should write the set of equations for this new situation.

As there is no motion of mass m_1 perpendicular to the plane, the forces acting perpendicular to the plane must be in equilibrium.

Resolving the forces on m_1 perpendicular to the plane, we obtain

$$R = m_1 g \cos \theta$$

Applying Newton's 2nd law of motion to mass m_1 we obtain

$$m_1 a = T - m_1 g \sin \theta$$

Applying Newton's 2nd law of motion to mass m_2 we obtain

$$m_2 a = m_2 g - T$$

Example

1   The diagram shows a particle A, on a **smooth** inclined plane, joined by a light inextensible string passing over a smooth pulley to a particle B, which hangs freely. The plane is inclined at an angle α to the horizontal, where $\sin \alpha = \frac{5}{13}$.

The masses of A and B are 13 kg and 15 kg respectively. The string is in the same vertical plane as a line of greatest slope of the plane.

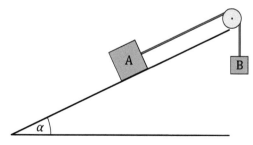

Initially, the particles are held at rest with the string taut.
The system is released. Calculate the magnitude of the acceleration of the particle A and the tension in the string.

As the surface of the slope is smooth, there is no frictional force acting. All the other forces are marked on the diagram. Notice that the weight of A (i.e. $13\,g$) can be resolved into two components; one $13\,g\sin\alpha$ parallel to the plane, the other $13\,g\cos\alpha$ at right angles to the plane.

Notice there are two unknowns in this equation (i.e. a and T).

Another equation is needed also containing the two unknowns so they can be found using simultaneous equations.

Answer

1   We will assume that when released particle A will accelerate up the slope as particle B moves vertically down.

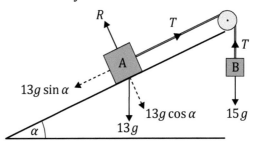

Applying Newton's 2nd law to particle A for motion up the plane, we have

$$ma = T - 13g\sin\alpha$$

$$13a = T - \left(13 \times 9.8 \times \frac{5}{13}\right)$$

$$13a = T - 49 \tag{1}$$

Applying Newton's 2nd law to particle B, we have

$$ma = 15g - T$$

$$15a = (15 \times 9.8) - T$$

$$15a = 147 - T \tag{2}$$

Solving equations (1) and (2) simultaneously gives $a = 3.5$ m s^{-2} and $T = 94.5$ N

7.3 Friction and the laws of friction

Friction

Friction is a force which opposes motion and occurs when two surfaces are in contact.

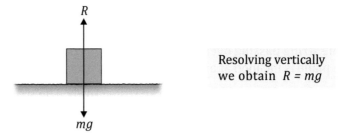

Resolving vertically we obtain $R = mg$

If the particle shown in the above diagram is stationary and there is no horizontal force acting, then the frictional force will be zero. The only forces acting will be the weight mg acting downwards, which will be in equilibrium with the normal reaction R acting upwards.

If a particle is subjected to a horizontal force T (see diagram below) but does not move, then the frictional force F must be equal to the horizontal force. The force of friction is equal and opposite to the horizontal force.

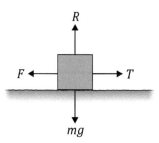

Resolving vertically we obtain
$$R = mg$$
Resolving horizontally we obtain
$$F = T$$
for equilibrium

If a particle is subjected to a horizontal force T that is larger than the frictional force F, then there will be a resultant force (i.e. $T - F$) which will cause the particle to accelerate. In the case shown above, the particle will move to the right with an acceleration a m s^{-2}.

Newton's 2nd law of motion can be applied and we obtain

$$ma = T - F$$

As $T > F$, there will be a resultant force to the right, which will produce an acceleration also to the right. Here, we use the direction to the right as the positive direction.

Laws of friction

There are a number of laws which can be applied to situations where friction occurs and these are summarised here:

- For two bodies in contact, the force of friction opposes the relative motion of the bodies.

- If bodies are in equilibrium, the force of friction is just sufficient to prevent motion and can be found by resolving the forces parallel to the surface.

- The size of the frictional force which can be exerted between two bodies is limited. If the force acting on a body is great enough, then motion will occur. Limiting friction is the frictional force exerted when equilibrium is on the point of being broken.

Example

1 A particle of mass 4 kg is pulled along a rough horizontal surface by a 20 N force. If the frictional force is 8 N, and the particle is initially at rest, calculate:

(a) The acceleration of the particle.

(b) The distance covered by the particle in the first 5 seconds.

Always check the question to see if a surface is rough or smooth. If a surface is smooth, then no frictional forces will act (i.e. $F = 0$).

Answer

1 (a) Applying Newton's 2nd law, we obtain
$$ma = 20 - 8$$
$$4a = 12$$
$$a = 3 \text{ m s}^{-2}$$

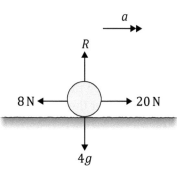

(b) $u = 0$, $a = 3$, $t = 5$, $s = ?$

Using $\quad s = ut + \frac{1}{2}at^2$

$$s = (0 \times 5) + \left(\frac{1}{2} \times 3 \times 5^2\right) = 37.5 \text{ m}$$

The resultant force is 20 – 8 to the right and this is used to accelerate the mass according to the equation,
$$F = ma.$$

Looking at the sizes of the horizontal forces you can see that the force to the right is greater than the force to the left. The resultant force and hence the acceleration will be to the right.

The equations of motion are used here. Remember that these will not be given so you need to memorise them.

7.4 Limiting friction and the coefficient of friction

When a particle is on a rough horizontal surface and an increasing horizontal force is applied, there comes a point when the particle will start to move. This is because the frictional force can increase to oppose the force trying to make it move only up to a certain amount. This amount is called the maximum or limiting friction.

This maximum or limiting value for the frictional force depends on the following two things:

- The size of the normal reaction between the two surfaces in contact.

- The roughness of the two surfaces in contact.

The roughness of the surfaces in contact is measured using a quantity called the coefficient of friction which is given the symbol, μ.

The maximum or limiting frictional force between two surfaces can be found using the following equation:

$$\text{Limiting friction, } F = \mu R$$

This equation can also be expressed in the following way

$$F_{MAX} = \mu R$$

where μ is the coefficient of friction and R is the normal reaction between the surfaces.

When a force P acts on a block of mass m on a rough, horizontal surface the maximum (or limiting) value that F can take is given by $F_{MAX} = \mu R$.

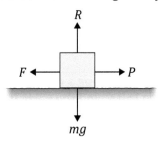

There are three situations that can occur as the force P varies and they are:

- When $P < F_{MAX}$ the frictional force F will be equal to P and as these forces balance, the block will be in equilibrium and remain at rest.

- When $P = F_{MAX}$ the frictional force has reached its maximum value. The block experiences limiting friction and is just on the point of moving.

- When $P > F_{MAX}$ then there will be a resultant force which will start to accelerate the block. Throughout the motion, the frictional force will be at its maximum value.

Examples

1   An object, of mass 5 kg, lies on a **rough**, horizontal surface. The coefficient of friction between the object and the surface is 0.6. A horizontal force of magnitude T N is applied to the object.

   (a)   Given that $T = 40$, calculate the magnitude of the frictional force and the acceleration of the object.

If you are dealing with an object on the point of moving or actually moving, then the object will experience the maximum frictional force.

BOOST

Grade

Many students get confused about the idea of limiting friction. Limiting friction is the maximum friction an object can experience. Friction is not a constant force. Instead it can have a value from zero up to a maximum called limiting friction.

(b) Given that $T = 20$, describes what happens, giving a reason for your answer.

Answer

1 (a)

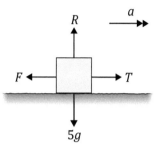

Resolving forces in the vertical direction, we obtain

$$R = 5g$$

$$R = 5 \times 9.8 = 49\,\text{N}$$

Limiting friction, $F = \mu R = 0.6 \times 49 = 29.4\,\text{N}$

Applying Newton's 2nd law in the horizontal direction, we have

$$ma = T - F$$

$$5a = 40 - 29.4$$

$$a = 2.12\,\text{m s}^{-2}$$

> Draw a diagram and mark on it all the forces acting on the object. Also mark the direction of the acceleration, which will be in the direction of the resultant force.
>
> Take this direction as the positive direction.

> The forces are balanced in the vertical direction, so the normal reaction, R, is equal and opposite to the weight $5g$.

(b) The particle will remain at rest.

T is 20 N which is less than the limiting friction of 29.4 N. The frictional force will therefore be equal to 20 N and there will be no resultant force, no acceleration and no motion.

2 A mass of 10 kg is at rest on a **rough**, horizontal surface. It is attached by a light string which passes over a smooth pulley to a mass of 5 kg which hangs vertically. The arrangement is shown in the following diagram.

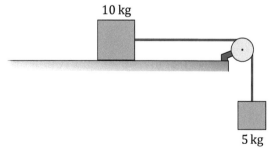

The coefficient of friction between the block and the surface is 0.2 and the system is released from rest and the block accelerates with an acceleration, a.

(a) Find the magnitude of the friction force acting on the 10 kg block.

(b) Find the acceleration of the system, a.

(c) Find the tension in the string.

BOOST

Grade

Draw a diagram showing all the forces acting on each block and also mark the direction of the acceleration of each block.

Remember that friction can take any value up to a certain maximum value which can be referred to either as limiting friction or F_{MAX}.

As the block is moving, it will experience the maximum friction.

BOOST

Grade

Do not make the mistake of including either the weight or the normal reaction in this equation. Both of these act only in the vertical direction and this equation of motion refers only to horizontal motion.

BOOST

Grade

Remember to check the values by substituting them into equation (2) and check that the left-hand side of the equation equals the right-hand side.

Answer

2   (a)

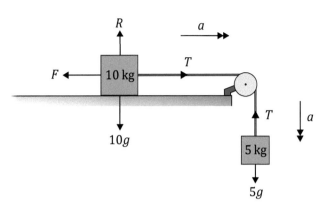

Resolving vertically for the 10 kg mass, we obtain

$$R = 10g = 98\,\text{N}$$

Limiting friction $F_{MAX} = \mu R = 0.2 \times 98 = 19.6\,\text{N}$

(b)   Resolving horizontally and applying Newton's 2nd law to the 10 kg mass, we obtain

$$ma = T - F$$

$$10a = T - 19.6 \tag{1}$$

Resolving vertically and applying Newton's 2nd law to the 5 kg mass, we obtain

$$5a = 5g - T$$

so $5a = 49 - T$ $\tag{2}$

Adding equations (1) and (2) we obtain

$$15a = 29.4$$

$$a = 1.96\,\text{m s}^{-2}$$

(c)   Substituting the value of a into equation (1) we obtain

$$19.6 = T - 19.6$$

$$T = 39.2\,\text{N}$$

---

3   The diagram shows an object, of mass 8 kg, on a **rough** plane inclined at an angle of 15° to the horizontal.

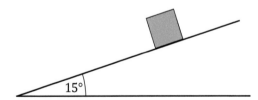

(a)   Given that the object is at rest, calculate the least possible value of the coefficient of friction. Give your answer correct to two decimal places.

(b)   Given that the coefficient of friction is 0.1, find the acceleration of the object down the plane.

Answer

3   (a)

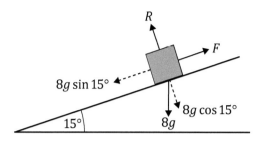

Component of the weight parallel to the plane = $8g \sin 15°$

Component of the weight at right angles to the plane = $8g \cos 15°$

(Note both these forces are marked on the diagram and we can disregard the weight as it has been replaced by its two components.)

The least value of the frictional force is when it just balances the component of the weight acting down the slope.

Hence, resolving forces parallel to the slope, we obtain

$$F = 8g \sin 15° = 8 \times 9.8 \sin 15° = 20.29 \text{ N}$$

Resolving perpendicular to the slope, we obtain

$$R = 8g \cos 15° = 75.73 \text{ N}$$

Limiting friction $F = \mu R$

Hence $\mu = \dfrac{F}{R} = \dfrac{20.29}{75.73} = 0.27$ (correct to two decimal places)

(b)   Limiting friction, $F = \mu R = 0.1 \times 75.73 = 7.57 \text{ N}$

Applying Newton's 2nd law, we obtain

$$ma = mg \sin 15° - F$$

$$8a = 8 \times 9.8 \sin 15° - 7.57$$

Hence $a = 1.59 \text{ m s}^{-2}$

Remember that when you deal with objects on slopes you need to use the components of the weight at right angles to the slope and parallel to the slope.

The component of the weight parallel to the slope will need a balancing force (i.e. friction in this case) if the object is to remain at rest.

BOOST

Grade

You must say that this is limiting friction or F_{MAX}.

This is the maximum friction that can occur and this value will be the friction acting on the block when it is moving.

---

4   The diagram shows two bodies A and B, of mass 9 kg and 5 kg respectively, connected by a light, inextensible string passing over a smooth, light pulley fixed at the edge of a **rough**, horizontal table. The heavier body A lies on the table and the lighter body B hangs freely below the pulley.

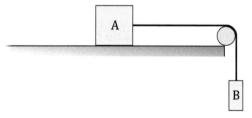

Initially, the system is held at rest with the string taut. The system is then released.

(a)   Given that the magnitude of the acceleration of the bodies is 1.61 m s^{-2}, calculate the tension in the string and the coefficient of friction between A and the table.

(b)   Given that the coefficient of friction is 0.6, determine whether the bodies will move or remain at rest and evaluate the tension in the string.

Answer

4   (a)

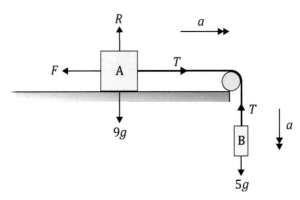

Applying Newton's 2nd law to B, we obtain:

$$5g - T = 5a$$

$$T = 5 \times 9.8 - 5 \times 1.61 = 40.95 \text{ N}$$

Resolving vertically for *A*, we obtain:

$$R = 9g = 88.2 \text{ N}$$

Now as A is moving, it experiences the limiting frictional force given by:

$$F = \mu R$$

$$= 88.2\mu$$

Applying Newton's 2nd law to *A*, we obtain:

$$9a = T - F$$

Hence, $9 \times 1.61 = 40.95 - 88.2\mu$

Solving, gives $\mu = 0.3$

(b)   Limiting friction, $F = \mu R = 0.6 \times 9g = 5.4g$

Applying Newton's 2nd law to the motion of B, we obtain:

$$5a = 5g - T$$

If the arrangement was at rest, $a = 0$ so $5g - T = 0$ and $T = 5g$

This limiting friction (5.4*g*) is greater than 5*g* so the particle will remain at rest.

If there is no acceleration, there is no resultant force, so $T = 5g$

$$= 49 \text{ N}$$

---

5   Two particles P and Q are connected by a light string. Particle P has a mass of 5 kg and lies on a **rough**, horizontal table. The string passes over a smooth pulley on the edge of the table and particle Q of mass 8 kg hangs vertically at the other end of the string. Particle Q is then released from rest at a height of 1.5 m above the ground and it takes a time of 1 s to hit the ground. Find the tension in the string, the acceleration of the particles and the frictional force acting on P.

Answer

5

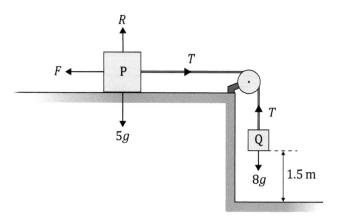

> This question is harder because no diagram is included. You need to draw your own and mark the forces and other information on it.

For particle Q, we have $u = 0$, $s = 1.5$, $t = 1$, $a = ?$

Using $\qquad s = ut + \dfrac{1}{2}at^2$

$$1.5 = 0 + \left(\dfrac{1}{2} \times a \times 1^2\right)$$

$$a = 3\ \mathrm{m\,s^{-2}}$$

> One of the equations of motion is used here to calculate the acceleration of the system when particle Q is released from rest.

Applying Newton's 2nd law to particle Q, we have

$$8a = 8g - T$$

$$8 \times 3 = (8 \times 9.8) - T$$

$$T = 54.4\ \mathrm{N}$$

Applying Newton's 2nd law to particle P, we have

$$ma = T - F$$

$$5 \times 3 = 54.4 - F$$

$$F = 39.4\ \mathrm{N}$$

---

6 The diagram shows two blocks, P with a mass of 5 kg and Q with a mass of 3 kg. Both blocks are connected by a light inextensible string passing over a smooth pulley. Block P is situated on a rough horizontal table and particle Q hangs vertically below the pulley.

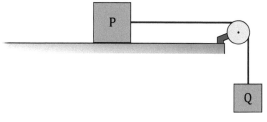

Initially, the system is held at rest with the string just taut. The system is then released.

(a) If the coefficient of friction between P and the table is 0.3, find the magnitude of the acceleration of particle P and the tension in the string.

(b) When released, the particles now remain stationary. Find the least value of the coefficient of friction for this to happen.

It is assumed that the tension is greater than the limiting friction so that P moves to the right.

Answer

6   (a)

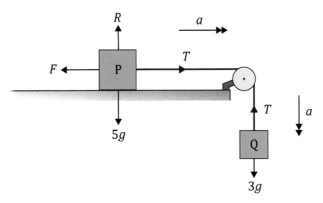

Resolving vertically for block P, we obtain

$$R = 5g$$

Limiting friction, $F = \mu R$

$$= 0.3 \times 5g$$

$$= 1.5g$$

Applying Newton's 2nd law to block P, we have

$$ma = T - F$$

$$5a = T - 1.5g \tag{1}$$

Applying Newton's 2nd law to block Q, we have

$$3a = 3g - T \tag{2}$$

Adding equations (1) and (2), we obtain

$$8a = 1.5g$$

$$a = 1.84 \text{ m s}^{-2}$$

Substituting the value of a into equation (1), we obtain

$$5 \times 1.84 = T - (1.5 \times 9.8)$$

$$T = 23.9 \text{ N}$$

Don't forget to check these two values (i.e. a and T) by substituting them back into equation (2). The left-hand side of the equation should equal the right-hand side.

(b) When the blocks are stationary, acceleration $a = 0$. This means that for each block, the forces are in equilibrium.

Resolving forces vertically for block P, we obtain

$$R = 5g$$

Resolving forces vertically for block Q, we obtain

$$T = 3g$$

Do not use the value of T as calculated in part (a) as the tension will be different when the particles are not accelerating.

Applying Newton's 2nd law of motion horizontally for block P, we obtain

$$ma = T - F$$

Now $F_{MAX} = \mu R = 5\mu g$ and $a = 0$, so

$$0 = 3g - 5\mu g$$

Here the values of F, a and T are substituted into the equation $ma = T - F$

Hence least value of $\mu = 0.6$

7 An object of mass 60 kg lies on a rough plane inclined at an angle of 25° to the horizontal. The coefficient of friction between the plane and the object is denoted by μ. Initially, the object is held at rest. It is then released.

(a) When $\mu = 0.3$, the object slides down the plane. Calculate:

(i) The magnitude of the frictional force.

(ii) The acceleration of the object.

(b) Given that when the object is released it remains stationary, calculate the least possible value of μ.

Answer

7 (a) (i)

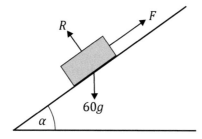

$R = 60g \cos \alpha$

$F = \mu R$

$= 60 \times 9.8 \cos \alpha \times 0.3$

$= 159.87 \, \text{N}$

If there are forces parallel or perpendicular to the slope then it is easier to solve these problems if the other forces are resolved in these directions.
Newton's 2nd law of motion can be applied to the motion parallel to the slope.

(ii) Applying Newton's 2nd law to the object, we obtain:

$60g \sin \alpha - F = 60a$

$60a = 60 \times 9.8 \sin 25° - 159.87$

$a = 1.4 \, \text{m s}^{-2}$

Taking the downwards direction as positive.

(b) As the object remains stationary, component of the weight acting down the slope must be equal to the frictional force.

At the point of slipping

$60g \sin \alpha = \mu \times 60g \cos \alpha$

The least value of $\mu = \dfrac{\sin \alpha}{\cos \alpha} = \tan \alpha$

$= \tan 25°$

$= 0.4663$

$= 0.47 \, (2 \, \text{d.p.})$

Divide both sides by $g \cos \alpha$.

Test yourself

1   A light inextensible string connects particles P of mass 5 kg and Q of mass 8 kg. Particle P lies on a smooth inclined plane which is inclined at an angle of 30° to the horizontal. The string passes over a smooth pulley and both particles are released from rest. The system is shown below.

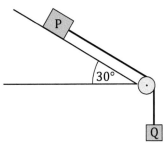

(a) Find the acceleration of the system.   [4]
(b) Find the tension in the string.   [2]

2   A horizontal force of 40 N acts on a block of mass 5 kg causing it to move in a straight line along a **rough** horizontal surface. The arrangement is shown in the diagram.

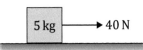

(a) Draw a diagram showing all the forces acting on the block.   [2]
(b) Find the magnitude of the normal reaction acting on the block.   [1]
(c) If the acceleration of the block is 2 m s^{-2}, calculate the magnitude of the frictional force acting on the block.   [2]

3   A particle of mass 3 kg lies at rest on a rough horizontal surface. The coefficient of friction between the particle and surface is 0.3.
(a) Find the magnitude of the frictional force and the acceleration of the particle, if a horizontal force P of magnitude 8 N is applied to the block.   [2]

(b) Find the magnitude of the frictional force and the acceleration of the particle, if a horizontal force P of magnitude 12 N is applied to the block.   [3]

4   A box of mass 6 kg rests on a rough plane inclined at an angle of 20° to the horizontal.

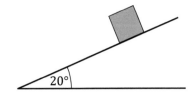

(a) Calculate the least value of the coefficient of friction, if the box is to remain at rest on the slope.   [5]
     Give your answer correct to two significant figures.

(b) If the coefficient of friction is 0.2, calculate the acceleration of the box down the slope.   [3]

5 A particle of mass 3 kg moves in a straight line on a rough, horizontal surface. The coefficient of friction between the particle and the surface is $\frac{6}{49}$.

(a) Find the frictional force and show that the deceleration of the particle is 1.2 m s^{-2}. [4]

(b) The speed of the particle at the point O is 9 m s^{-1} and it comes to rest at point A. Calculate the distance OA. [3]

6 A sledge, of mass 39 kg, moves on a rough slope inclined at an angle to the horizontal, where $\tan \alpha = \frac{5}{12}$. The coefficient of friction between the sledge and the slope is 0.3.

(a) Given that the sledge is moving freely down a line of greatest slope, calculate the magnitude of the acceleration of the sledge. Give your answer correct to two decimal places. [6]

(b) Given that the sledge is being pulled up the slope with acceleration 0.4 m s^{-2} by means of a rope parallel to a line of greatest slope, find the tension in the rope. [3]

Summary

Check you know the following facts:

Newton's laws of motion

1st law – a particle will remain at rest or will continue to move with constant speed in a straight line unless acted upon by some external force.

2nd law – a resultant force produces an acceleration, according to the formula

Force = mass × acceleration.

3rd law – every action has an equal and opposite reaction.

Finding a resultant force or acceleration

Resolve all the forces in two perpendicular directions and then use Pythagoras' theorem to find the resultant of these two forces at right angles. Use trigonometry to find the angle and clearly state the direction.

Use $a = \dfrac{F}{m}$ to find the acceleration which will be in the direction of the resultant force.

Motion on an inclined plane

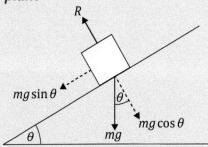

The component of the weight parallel to the slope = $mg\sin\theta$

The component of the weight at right angles to the slope = $mg\cos\theta$

Friction opposes motion

It can increase up to a certain maximum value called limiting friction or F_{MAX}.

Limiting friction, $F = \mu R$ which can also be written as $F_{MAX} = \mu R$, where μ is the coefficient of friction and R is the normal reaction between the surfaces.

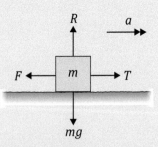

If $T > F$ there will be an unbalanced force and the mass will accelerate.

By Newton's 2nd law $ma = T - F$

Once moving, the mass will experience the maximum frictional force given by

$$F_{MAX} = \mu R$$

The forces are balanced in the vertical direction, so $R = mg$

8 Projectile motion

Introduction

In this topic you will be modelling motion in a vertical plane under the effect of gravity using vectors. When an object is projected at a certain velocity in the vertical plane, the object is called a projectile and the subsequent motion is referred to as projectile motion. As a projectile is just given an initial velocity, it is not powered, so examples of a projectile include a dart or javelin being thrown, a golf ball being hit, a cannon being fired, a ball being thrown, etc.

As with most mechanics systems, we have to make some assumptions in order to simplify the mathematics. These modelling assumptions are important and you will be asked about them in the examination.

This topic covers the following:

8.1   Modelling assumptions

8.2   Projectiles in the horizontal direction

8.3   Motion of a projectile projected at an angle to the horizontal

8.1 Modelling assumptions

As always in mechanics, real life is complicated, so we make some modelling assumptions to keep the mathematics simple. For projectile motion, the modelling assumptions are:

1   The body (stone, ball, bullet, etc.) can be treated as a particle. This will eliminate spin, which could deflect the body out of the vertical plane.

2   Friction is assumed to be zero, so the only force acting on the particle will be the gravitational force.

3   The acceleration due to gravity, g, is constant.

4   The motion is confined to the vertical plane.

8.2 Projectiles in the horizontal direction

Suppose a ball is projected horizontally from a cliff which is 20 m above sea level with a velocity of 8 m s^{-1}. We can start off by sketching this arrangement:

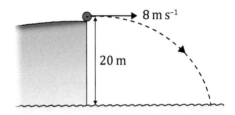

The motion of the projectile is affected by gravity, which acts only in the vertical direction. The particle follows a parabolic path as shown in the diagram.

Since gravity only acts in the vertical direction, it has no effect on the horizontal velocity, which remains constant at 8 m s^{-1}.

The vertical motion is affected by the gravity which acts in the vertical direction and the vertical velocity of the projectile is 0 m s^{-1} at the start and then accelerates to the vertical velocity with which it hits the water.

It is important to note that the vertical and horizontal motions can be considered separately.

The following equations of motion are used in this topic:

$$v = u + at$$

$$s = ut + \frac{1}{2}at^2$$

$$v^2 = u^2 + 2as$$

$$s = \frac{1}{2}(u + v)t$$

s = displacement/distance
u = initial velocity/speed
v = final velocity/speed
a = acceleration
t = time

For the horizontal motion, as there is no acceleration in this direction, we use the equation

$$\text{Speed} = \frac{\text{distance travelled}}{\text{time taken}}$$

Finding the time of flight for a particle projected horizontally

The time of flight is the time the projectile remains in the air, and this can be found by considering the vertical motion. It is important to note that the time the projectile spends in the air is the same as if it were simply dropped vertically from rest.

We need to first establish the direction we are regarding as positive. In this case we will take downwards as positive.

Considering the vertical direction, we have

$u = 0 \text{ m s}^{-1}$ (this is the velocity in the vertical direction)

$a = g = 9.8 \text{ m s}^{-2}$

$s = 20 \text{ m}$ (i.e. the height of the cliff)

$t = ?$ (this is the required time of flight)

We can use the following equation of motion:

$$s = ut + \frac{1}{2}at^2$$

$$20 = 0 + \frac{1}{2} \times 9.8t^2$$

Solving gives $t = 2.0 \text{ s}$ (2 s.f.)

> The time taken is the time of flight (i.e. the time taken for the particle to fall vertically from rest).

Finding the range (i.e. the distance travelled in the horizontal direction) for a particle projected horizontally

As the velocity in the horizontal direction is constant we can simply use a rearrangement of

$$\text{Velocity} = \frac{\text{distance travelled}}{\text{time taken}} \quad \text{to find the range.}$$

So,   Range = velocity × time of flight = 8 × 2.0 = 16 m

> Note that the horizontal velocity is unaffected by gravity, which only affects the vertical velocity.

Examples

1   A particle is projected horizontally from a point that is 30 m above a horizontal surface. It hits the surface at a point that is 100 m in a horizontal direction from the point of projection. Find the initial speed of the projected particle.

. . . .

Answer

1   Let $u \text{ m s}^{-1}$ be the horizontal velocity the particle is projected with.

Finding the time of flight and taking the downward direction as positive, we have

$$s = ut + \frac{1}{2}at^2$$

$$30 = 0 + \frac{1}{2} \times 9.8t^2$$

Solving gives $t = 2.4744 = 2.5 \text{ s}$ (2 s.f.)

Range = velocity × time of flight

100 = velocity × time of flight

100 = velocity × 2.4744

Velocity = 40.4138 = 40 m s^{-1} (2 s.f.)

> Note that this time is the time of flight (i.e. the time spent in the air by the particle).

2 A steel ball is projected horizontally from a vertical cliff with a velocity of $100\,\text{m s}^{-1}$.

If the point of projection is 40 m above sea level, find the speed with which the ball hits the water.

Answer

2 Considering the vertical velocity, we have:

$$v^2 = u^2 + 2as$$

$$v^2 = 0^2 + 2 \times 9.8 \times 40$$

$$v = 28\,\text{m s}^{-1}$$

Now the horizontal velocity stays constant at $100\,\text{m s}^{-1}$.

Hence we can draw the following diagram showing the vertical and horizontal velocities:

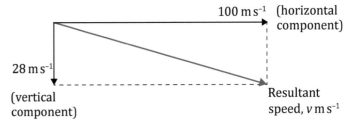

To find the resultant of these we use Pythagoras' theorem.

$$v^2 = 100^2 + 28^2$$

Giving $v = 104\,\text{m s}^{-1}$

8.3 Motion of a particle projected at an angle α to the horizontal

Here we will be looking at the motion of a particle projected with a speed of $U\,\text{m s}^{-1}$ and at an angle of α to the horizontal.

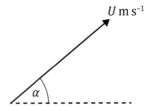

This velocity can be represented by vertical and horizontal components as shown below:

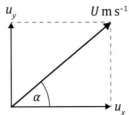

The horizontal and vertical components of the velocity U are u_x and u_y respectively.

By using the following right-angled triangle we can write these components in terms of u:

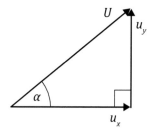

The horizontal component, $u_x = U \cos \alpha$

The vertical component, $u_y = U \sin \alpha$

These two equations are used in conjunction with the equations of motion with either the vertical or horizontal motions.

Finding the time of flight

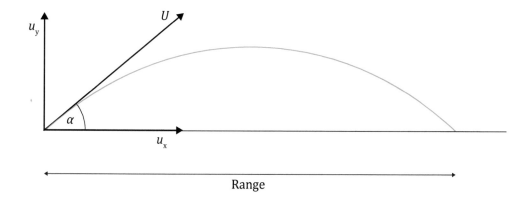

Range

To find the time of flight, we consider the vertical motion and equate the vertical displacement to zero. We will consider the upward direction as the positive direction for the motion.

$$s = ut + \frac{1}{2}at^2$$

$$s = U \sin \alpha\, t - \frac{1}{2}gt^2$$

Note that g is negative as it acts downwards.

As $s = 0$, $\quad 0 = U \sin \alpha\, t - \frac{1}{2}gt^2$

Hence $t\left(U \sin \alpha - \frac{1}{2}gt\right) = 0$ so either $t = 0$ or $U \sin \alpha - \frac{1}{2}gt = 0$.

The $t = 0$, is when the particle is projected, so we use the other solution.

You must always derive the equation before using it in the examination otherwise you will lose marks.

Hence, time of flight, $t = \dfrac{2U \sin \alpha}{g}$

Finding the range (i.e. the horizontal distance travelled)

As the horizontal velocity remains constant during the time of flight, to find the range we use:

$$\text{Range} = \text{horizontal velocity} \times \text{time of flight}$$

Now the horizontal velocity $= U\cos\alpha$ and time of flight $= \dfrac{2U\sin\alpha}{g}$ so substituting these into the above formula, we obtain

$$\text{Range} = U\cos\alpha\,\frac{2U\sin\alpha}{g}$$

$$= \frac{2U^2\sin\alpha\cos\alpha}{g}$$

Now $2\sin\alpha\cos\alpha \equiv \sin 2\alpha$

So
$$\text{Range} = \frac{U^2\sin 2\alpha}{g}$$

Finding the greatest height reached

At the greatest height the vertical component of the velocity is zero.

Considering the vertical motion with the upward direction as positive, we have:

$v = 0\ \text{m s}^{-1},\ u = U\sin\alpha,\ a = g = -9.8\ \text{m s}^{-2}$ and $s = ?$

$$v^2 = u^2 + 2as$$

$$0^2 = U^2\sin^2\alpha - 2gs$$

Maximum height $s = \dfrac{U^2\sin^2\alpha}{2g}$

Examples

1 A particle is projected from a point on a horizontal plane at an angle of 60° to the horizontal and with a speed of 32 m s^{-1}.

Find the greatest height reached giving your answer to two significant figures.

Answer

1

Considering the vertical motion with the upward direction as positive, we have:

$v = 0\ \text{m s}^{-1},\ u = 32\sin 60°,\ a = g = -9.8\ \text{m s}^{-2}$ and $s = ?$

$$v^2 = u^2 + 2as$$

$$0^2 = 768 - 2 \times 9.8s$$

Hence the maximum height $s = 39$ m (2 s.f.)

2 A body is projected at time $t = 0\,\text{s}$ from a point O with speed $V\,\text{m s}^{-1}$ in a direction inclined at an angle of θ to the horizontal.

(a) Write down expressions for the horizontal and vertical components x m and y m of its displacement from O at time t s.

(b) Show that the range R m on a horizontal plane through the point of projection is given by $\frac{V^2}{g}\sin 2\theta$

(c) Given that the maximum range is 392 m, find, correct to one decimal place:

   (i) the speed of projection,

   (ii) the time of flight,

   (iii) the maximum height attained.

· ·

Answer

2 (a) As the horizontal component of velocity stays constant,

      distance = velocity × time.

   Hence $x = (V\cos\theta)t$

   Considering the vertical motion with the upward direction as positive.

   Using $s = ut + \frac{1}{2}at^2$ we have $y = (V\sin\theta)t - \frac{1}{2}gt^2$

> Remember that g acts in the opposite direction to the direction we have set as positive so in this equation it is $-g$.

(b) To find the time of flight we let $y = 0$.

   Hence $(V\sin\theta)t - \frac{1}{2}gt^2 = 0$ and solving for gives $t = \dfrac{2V\sin\theta}{g}$

   Substituting this into the equation for x we obtain:

$$x = \text{Range} = (V\cos\theta)\left(\frac{2V\sin\theta}{g}\right)$$

   Now, $2\sin\theta\cos\theta = \sin 2\theta$

   Hence, $\text{Range} = \dfrac{V^2\sin 2\theta}{g}$

> Remember this from your trigonometry or you can use $\sin(A + B)$ from the formula booklet with both A and B equal to θ.

(c) (i) The maximum range occurs when $\sin 2\theta = 1$

     Hence, maximum range $= \dfrac{V^2}{g}$ and $392 = \dfrac{V^2}{9.8}$

     Giving $V = 62.0\,\text{m s}^{-1}$ (1 d.p.)

   (ii) Time of flight, $t = \dfrac{2V\sin\theta}{g}$

     Now $\sin 2\theta = 1$, $2\theta = \sin^{-1}1$, hence $2\theta = 90°$ so $\theta = 45°$

     Hence $t = \dfrac{2V\sin\theta}{g} = \dfrac{2 \times 62.0 \times \sin 45°}{9.8} = 8.947 = 8.9\,\text{s}$ (1 d.p.)

> **BOOST**
>
> **Grade** ⇧⇧⇧⇧
>
> Always look back at any equations you have been asked to prove as you will probably have to use them in the next parts of the question.

   (iii) The time to reach the maximum height is half the time of flight.

     Hence time to reach maximum height = 4.47 s

$$y_{max} = (V\sin\theta)t - \frac{1}{2}gt^2 = 62.0 \times \sin 45° \times 4.47 - \frac{1}{2} \times 9.8 \times 4.47^2$$

$$= 98.1\,\text{m} \text{ (1 d.p.)}$$

> You should remember this fact.

3   A player kicks a ball from a point A on horizontal ground so that 2.5 seconds later the ball just clears a bar at a point B. The point B is 3 m above the ground. The horizontal distance of B from A is 42 m.

(a) Calculate the horizontal and vertical components of the initial velocity of the ball.

(b) Find the magnitude of the velocity of the ball and the angle that the direction of the velocity makes with the horizontal as it passes the point B.

(c) Determine the horizontal distance from B to the point where the ball first hits the ground again.

. . . .

Answer

3   (a)

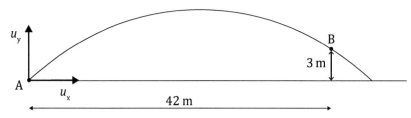

For the horizontal motion, $u_x = \dfrac{\text{distance}}{\text{time}} = \dfrac{42}{2.5} = 16.8 \text{ m s}^{-1}$

For the vertical motion, taking upwards as positive, we have:

$$s = ut + \frac{1}{2}at^2 \text{ so } 3 = u_y \times 2.5 - \frac{1}{2} \times 9.8 \times 2.5^2$$

Hence $u_y = 13.45 \text{ m s}^{-1}$

(b) For the horizontal motion $v_x = u_x = 16.8 \text{ m s}^{-1}$

For the vertical motion we use

$$v = u + at$$

so $v_y = u_y - gt = 13.45 - 9.8 \times 2.5 = -11.05 \text{ m s}^{-1}$

We can show these two components in a diagram along with actual velocity:

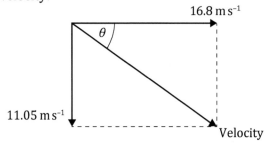

By Pythagoras' theorem we obtain:

Velocity² = 16.8² + 11.05²

Velocity = 20.11 m s⁻¹

$$\theta = \tan^{-1}\left(\frac{11.05}{16.8}\right)$$

$\theta = 33.33°$ (below the horizontal as shown on the diagram)

Remember the horizontal component of velocity stays constant.

The negative velocity means the velocity is in the opposite direction to that taken as positive (i.e. upwards). Hence this velocity is in the downward direction.

(c) Considering the vertical component and letting the upward direction be positive, we have

$$s = ut + \frac{1}{2}at^2$$

At the range, $s = 0$ so

$$0 = 13.45t - \frac{1}{2} \times 9.8 \times t^2$$

Solving, gives $t = 2.7449$ s

$$\text{Range} = \text{horizontal velocity} \times \text{time of flight}$$
$$= 16.8 \times 2.7449$$
$$= 46.11 \text{ m}$$

Required distance $= 46.11 - 42 = 4.11$ m

Note that t will be the time of flight.

The equation for the path of a projectile projected at an angle α to the horizontal with a speed U

The equation of a projectile projected a speed u and angle α can be found in terms of the horizontal distance travelled x and the vertical distance travelled y in the following way:

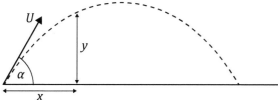

The horizontal component, $u_x = U \cos \alpha$

The vertical component, $u_y = U \sin \alpha$

As the horizontal component of velocity stays constant, distance = velocity × time.

Hence $x = (U \cos \alpha)t$,   so $t = \dfrac{x}{U \cos \alpha}$

Note that the equation of the projectile applies to any point on its path.

The vertical component of velocity is affected by gravity and taking upwards as the positive direction, we have

$$s = ut + \frac{1}{2}at^2$$

$$y = U \sin \alpha t - \frac{1}{2}gt^2$$

Note that g is negative as it acts downwards.

Now substituting $t = \dfrac{x}{U \cos \alpha}$ into the above equation, we obtain:

$$y = U \sin \alpha \frac{x}{U \cos \alpha} - \frac{1}{2}g \left(\frac{x}{U \cos \alpha} \right)^2$$

$$y = x \tan \alpha - \frac{gx^2}{2U^2 \cos^2 \alpha}$$

$$y = x \tan \alpha - \frac{gx^2 \sec^2 \alpha}{2U^2}$$

$\dfrac{\sin \alpha}{\cos \alpha} = \tan \alpha$

$\dfrac{1}{\cos \alpha} = \sec \alpha,$

so $\dfrac{1}{\cos^2 \alpha} = \sec^2 \alpha$

Now, as $\sec^2 \alpha = 1 + \tan^2 \alpha$, $y = x \tan \alpha - \dfrac{gx^2 (1 + \tan^2 \alpha)}{2U^2}$

You must remember how to derive this equation and not simply remember the result.

Step by STEP

A and B are points a distance 18 m apart on horizontal ground. An object P is projected from A towards B with velocity 15 m s⁻¹ at an angle of 60° to the horizontal. Simultaneously, another object Q is projected from B towards A with velocity v m s⁻¹ at an angle of 30° to the horizontal. The objects collide.

(a) Find the value of v.

(b) Show that the time from projection to collision is 0.6 seconds.

(c) Determine the speed of the object P just before collision.

Steps to take

1   First think about each projectile separately and what you know. As this is all about the colliding particles, think about the horizontal and vertical distances for each particle.

    For a collision, the vertical distances for each particle would have to be the same at the same time.

    However, the horizontal distances will probably be different for each particle but the total of these two distances will equal the distance between the points of projection of the particles.

2   Find the vertical component of each particle's velocity.

    Use the equation of motion $s = ut + \frac{1}{2}at^2$ to find the y-distance travelled by each particle. Remember to substitute the vertical component of the velocity into the formula for u.

3   Equate the equations representing the vertical distance travelled by each particle.

    Solve the resulting equation to find the value of v.

4   Find the horizontal components of the velocity for each particle. If the particles collide at time t, we can write two expressions for the distance travelled in the horizontal direction in this time. If we add these distances together we can equate them to the distance between the points of projection which is given in the question.

5   To determine the speed of P just before collision we use the initial vertical component as u and then use $v = u + at$ to work out its vertical velocity before collision. As the horizontal component remains constant we can use this with the vertical velocity and apply Pythagoras' theorem to determine the speed of P.

. . . .

Answer

(a)   Initial vertical velocity of P = 15 sin 60° = $\dfrac{15\sqrt{3}}{2}$ = 12.99 m s⁻¹

Initial vertical velocity of Q = v sin 30°

Let t be the time when both objects collide and let the upward direction be positive.

Vertical height of P at collision, $y = ut + \frac{1}{2}at^2$

$$= \frac{15\sqrt{3}}{2}t - 0.5gt^2$$

Vertical height of Q at collision, $\quad y = ut + \frac{1}{2}at^2$

$$= (v\sin 30°)t - \frac{1}{2}gt^2$$

$$= 0.5vt - 0.5gt^2$$

> Note sin 30° = 0.5

When the particles collide, the vertical distances for each particle will be equal.

Equating the vertical distances, we obtain

$$\frac{15\sqrt{3}}{2}t - 0.5gt^2 = 0.5vt - 0.5gt^2$$

> Dividing both sides by t.

$$\frac{15\sqrt{3}}{2} = 0.5v$$

$$v = 15\sqrt{3} = 25.98 \text{ m s}^{-1}$$

(b) Initial horizontal velocity of P = 15 cos 60°= 7.5 m s^{-1}

Initial horizontal velocity of Q = $15\sqrt{3}$ cos 30°= 22.5 m s^{-1}

> Remember that t is the time each particle is in the air before colliding.

Distance travelled horizontally by P to point of collision = $7.5t$

Distance travelled horizontally by Q to point of collision = $22.5t$

At the point of collision both distances have to equal 18 m, so

$$7.5t + 22.5t = 18$$

Solving, gives $\qquad\qquad t = 0.6$ s

(c) Using $v = u + at$ to work out the vertical velocity of P before collision.

$$v = u + at = \frac{15\sqrt{3}}{2} - 9.8 \times 0.6 = 7.1 \text{ m s}^{-1}$$

The horizontal velocity of P is constant at 7.5 m s^{-1}

Using Pythagoras' theorem, speed = $\sqrt{7.1^2 + 7.5^2}$

$$= 10.3 \text{ m s}^{-1}$$

Using vectors to describe projectile motion

Vectors can be used to describe projectile motion in the vertical plane with the **i** unit vector directed in the horizontal direction and the **j** unit vector directed in the vertical direction.

The following example shows how unit vectors can be used.

Examples

1 A particle is projected with an initial velocity of $(5\mathbf{i} + 12\mathbf{j})$ m s^{-1}.

Find:

(a) The initial speed of projection.

(b) The angle to the horizontal of projection to the nearest 0.1°.

Answer

1 (a)

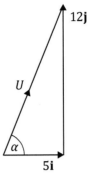

By Pythagoras' theorem, $\quad U^2 = 12^2 + 5^2$

$$= 169$$

$$U = 13 \text{ m s}^{-1}$$

(b) $\quad \tan \alpha = \dfrac{12}{5}$

$$\alpha = \tan^{-1}\left(\dfrac{12}{5}\right)$$

$$= 67.4° \text{ (to nearest } 0.1°)$$

---

2 A body is projected from a point O with velocity $(12\mathbf{i} + 5\mathbf{j})$ m s^{-1}.

(a) Find the position vector of the body after 2 s.

(b) Find the speed of the body after 2 s.

Answer

2 (a) Using the equation of motion, $s = ut + \dfrac{1}{2}at^2$ we can write the following equation in terms of the vectors:

$$\mathbf{s} = \mathbf{u}t + \dfrac{1}{2}\mathbf{a}t^2$$

$$\mathbf{s} = (12\mathbf{i} + 5\mathbf{j})2 - \dfrac{1}{2} \times 9.8\mathbf{j} \times 2^2$$

s is the displacement vector which is relative to the starting position which in this case is the origin.

$$\mathbf{s} = 24\mathbf{i} + 10\mathbf{j} - 19.6\mathbf{j}$$

$$\mathbf{s} = 24\mathbf{i} - 9.6\mathbf{j}$$

As this displacement vector is relative to the origin O, this will be the same as the position vector.

Hence position vector relative to O is $\mathbf{r} = 24\mathbf{i} - 9.6\mathbf{j}$

(b) Using the equation of motion $v = u + at$ we can write:

$$\mathbf{v} = \mathbf{u} + \mathbf{a}t$$

$$\mathbf{v} = (12\mathbf{i} + 5\mathbf{j}) - 9.8\mathbf{j} \times 2$$

$$\mathbf{v} = 12\mathbf{i} - 14.6\mathbf{j}$$

Using Pythagoras' theorem, speed $= \sqrt{12^2 + (-14.6)^2}$

$$= 18.9 \text{ m s}^{-1}$$

This equation of motion will give the displacement vector **s** relative to the starting position, which in this case is O. Notice that the acceleration due to gravity is taken as $-9.8\mathbf{j}$ m s^{-2}.

Active Learning

You should not use the formulae in this topic without first deriving them from first principles.

Produce a sheet with pointers for the derivations for the following formulae for a particle projected with a velocity U m s^{-1} and inclined at an angle α to the horizontal:

- Finding the time of flight,
- Finding the range,
- Finding the maximum height,
- Finding the equation of the path of the particle.

Test yourself

1. A sea eagle catches a fish. When it is flying in a horizontal direction with a speed of 6 m s^{-1} it drops the fish. The fish falls a vertical distance of 10 m into the sea below. The motion of the fish is to be modelled mathematically as a particle.
   (a) State two modelling assumptions you will make in order to model the motion of the fish after being released. [2]
   (b) Calculate the time between the fish being dropped by the eagle and it reaching the surface of the sea. Give your answer to two decimal places. [2]
   (c) Calculate the horizontal distance travelled by the fish from when it is dropped to when it reaches the surface of the sea. [1]

2. A stunt motorcyclist is attempting to clear a canyon which is 75 m wide. He uses a ramp which is inclined at 30° to the horizontal and takes off from this ramp with a speed of 30 m s^{-1}. Work out, by calculations, whether he clears the canyon or not. [6]

3. A particle P is projected from the origin O with velocity $(8\mathbf{i} + 17\mathbf{j}) \text{ m s}^{-1}$ and moves freely under gravity.
   Find:
   (a) the position vector of P in terms of t, [3]
   (b) the greatest height reached by the particle. [3]

4. A particle is projected from horizontal ground with speed 24.5 m s^{-1} in a direction inclined at an angle of 30° above the horizontal.
   (a) Calculate the horizontal range of the particle. [6]
   (b) Determine the maximum height reached by the particle. [3]
   (c) Write down the speed and the direction of motion of the particle as it hits the ground. [3]

5. A pebble is projected from a point A which is 5.4 m vertically above a point O on horizontal ground.

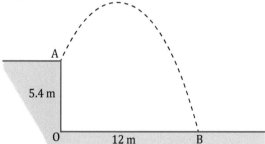

   The initial velocity of the pebble is $V \text{ m s}^{-1}$ at an angle α above the horizontal, where $\tan \alpha = \frac{3}{4}$. The pebble hits the ground at the point B which is at a distance of 12 m from O.
   The time of flight of the pebble is T s.
   (a) Write down the horizontal component and the vertical component of the initial velocity of the pebble in terms of V. [2]
   (b) Show that $VT = 15$. [2]
   (c) Find the value of T and hence find the value of V. [4]
   (d) Determine the speed of the pebble as it hits the ground at B. [5]

Summary

Check you know the following facts:

The equations of motion (all of these must be memorised)
The following equations of motion are used in this topic:

$$v = u + at$$

$$s = ut + \frac{1}{2}at^2$$

$$v^2 = u^2 + 2as$$

$$s = \frac{1}{2}(u + v)t$$

s = displacement/distance
u = initial velocity/speed
v = final velocity/speed
a = acceleration
t = time

If the acceleration is zero i.e. there is constant velocity/speed,

$$\text{speed} = \frac{\text{distance travelled}}{\text{time taken}}$$

Motion of a particle projected at an angle α to the horizontal with velocity $U\,m\,s^{-1}$

> Don't waste time trying to remember these formulae as you always need to derive them from first principles before using them.

The horizontal component, $u_x = U\cos\alpha$

The vertical component, $u_y = U\sin\alpha$

$$\text{Time of flight} = \frac{2U\sin\alpha}{g}$$

$$\text{Range} = \frac{U^2\sin 2\alpha}{g}$$

$$\text{Maximum height} = \frac{U^2\sin^2\alpha}{2g}$$

Equation for the path of a projectile at any point (x, y) along its path is given by

$$y = x\tan\alpha - \frac{gx^2(1 + \tan^2\alpha)}{2U^2}$$

9 Differential equations

Introduction

Simple differential equations were covered in A2 Unit 3 Pure so you need to make sure you fully understand the material in that before starting this topic.

This topic extends to the construction of differential equations to different contexts such as exponential growth and decay, kinematics, population growth and modelling the relationship between price and demand.

9.1 Formation of simple differential equations and their solution

Differential equations are equations concerning rate of change. Many problems in the real world involve quantities that change with time and therefore involve rates of change. For example, radioactive decay, the depreciation in value of a car, the growth of bacteria, all involve rates of change. By forming a differential equation, the situation can be accurately modelled and the model can be used to provide answers to questions such as, 'what will be the likely value of my car in four years' time?' or 'how much of a radioactive isotope will remain after a certain period of time?'

Active Learning

During this topic you will come across problems where a body moves through a fluid (a fluid is a gas or liquid) and experiences a resistive force which depends on the velocity in some way.
Do some research using the Internet to produce an explanation as to why the resistive force on the body increases as the velocity of the body increases.

9.2 Exponential growth and decay

Many quantities exhibit exponential growth or exponential decay as these examples show.

Examples

1 A radioactive substance decays so that at any instant, the rate of decrease in the mass is proportional to the mass of the radioactive substance remaining. If the mass of the radioactive substance is m grams after t days, write down a differential equation relating m and t.

When the proportional sign is changed to an equals sign a constant of proportionality k must be included. The minus sign is included here because the mass is decreasing (i.e. m decreases with time).

Answer

1   $\dfrac{dm}{dt} = -km$

2 The size N of the population of a small island may be modelled as a continuous variable. At time t, the rate of increase of N is directly proportional to the value of N.

(a) Write down the differential equation that is satisfied by N.

(b) Show that $N = Ae^{kt}$, where A and k are constants.

(c) Given that $N = 100$ when $t = 2$ and that $N = 160$ when $t = 12$,

   (i) show that $k = 0.047$, correct to three decimal places,

   (ii) find the size of the population when $t = 20$.

Answer

When the proportionality sign is removed and replaced with an equals sign, a constant of proportionality k is included.
Since N increases,
$\dfrac{dN}{dt} > 0$ and $k > 0$.

2   (a) $\dfrac{dN}{dt} \propto N$

so that $\dfrac{dN}{dt} = kN$

This represents the phrase 'rate of increase of N is directly proportional to the value of N'.

(b) Separating the variables and integrating, we obtain

$$\int \frac{1}{N}\, dN = k \int dt$$

$$\therefore \ln N = kt + C \qquad\qquad (1)$$

If $N = N_0$ when $t = 0$,

$$\ln N_0 = k(0) + C = C$$

Substitute for C in (1)

$$\ln N = kt + \ln N_0$$

$$\therefore \ln N - \ln N_0 = kt$$

> Use one of the laws of logarithms
>
> $$\ln a - \ln b = \ln \frac{a}{b}$$

$$\ln \frac{N}{N_0} = kt$$

Taking exponentials of both sides

$$\frac{N}{N_0} = e^{kt}$$

$$\therefore N = N_0 e^{kt}$$

Writing $A = N_0$, we have $N = Ae^{kt}$

(c) (i) $N = 100$ when $t = 2$ and $N = 160$ when $t = 12$,

Hence $100 = Ae^{2k}$ and $160 = Ae^{12k}$

Dividing these two equations to eliminate A we obtain:

$$\frac{160}{100} = \frac{e^{12k}}{e^{2k}}$$

$$1.6 = e^{10k}$$

> Note $\dfrac{e^{12k}}{e^{2k}} = e^{12k-2k} = e^{10k}$

Taking ln of both sides we obtain:

$\ln 1.6 = 10k$ so $k = \dfrac{1}{10}\ln 1.6 = 0.047$ (correct to three decimal places)

(ii) $N = Ae^{0.047t}$

$N = 100$ when $t = 2$

so $\qquad 100 = Ae^{0.047 \times 2}$

$\qquad\qquad 100 = A \times 1.09856$

giving $\qquad A = 91.028$ (correct to three decimal places)

Hence $\qquad N = 91.028 e^{0.047t}$

When $t = 20$, $N = 91.028 e^{0.047 \times 20} = 233$ (nearest whole number)

Step by STEP

The rate of change of population of sea birds on an island is proportional to the size of the population P, with constant of proportionality k.
At time $t = 0$ hours, the size of the population is 720.

(a) Find an expression, in terms of k, for P at time t

(b) Given that the population increases by 1% after 48 hours, how long will it take for the population to increase by 5% of its original size?

Steps to take

1   In the question you are told that the rate of change of the population with respect to time is proportional to the population, so we first write this mathematically and then as an equation by including a constant of proportionality.

2   Now you have a differential equation for P but this is not what the question asked for. The question asked for an expression for P in terms of k at a time t, therefore we must separate the variables and integrate.

3   After the integration, a constant of integration will need to be included so it is necessary to find a value for this constant. We use the information when $t = 0$ hours, the size of the population is 720 to enter these values and hence find the value of the constant.

4   The resulting expression is then rearranged and simplified.

5   For part (b) we use the increased population along with the time of 48 hours to find a value for the constant k.

6   We can then substitute the value of k back in and then use the resulting equation to find the time when the population has increased by 5%.

· ·

Answer

(a) $$\frac{dP}{dt} \propto P$$

$$\frac{dP}{dt} = kP$$

Separating the variables and integrating, we obtain:

$$\int \frac{dP}{P} = \int k\,dt$$

$$\ln P = kt + C$$

$$\ln 720 = C$$

Hence, $\ln P = kt + \ln 720$

$$\ln \frac{P}{720} = kt$$

> Now you know that when $t = 0$ hours $P = 720$, so from this we can find the value for the constant of integration, C.

Taking exponentials of both sides, we obtain:

$$\frac{P}{720} = e^{kt}$$

$$P = 720e^{kt}$$

(b)   Using the values we are given in the question we can find a value for the constant k when $t = 48$ and $P = 727.2$ (the original population plus the 1% increase).

$$727.2 = 720e^{48k}$$

$$1.01 = e^{48k}$$

Taking ln of both sides, we obtain:

$$\ln(1.01) = 48k$$

$$k = 0.0002073$$

New population after 5% rise $= 1.05 \times 720 = 756$

$$756 = 720e^{0.0002073t}$$

$$1.05 = e^{0.0002073t}$$

Taking ln of both sides, we obtain:

$$\ln 1.05 = 0.0002073t$$

$$t = 235 \text{ hours}$$

Active Learning

You may have to integrate a variety of expressions in this topic. Produce a crib sheet of the various methods of integration that you have covered in other parts of the course.

Examples

1 Floating pond weed in a pond covers an area of A m² after time t months. The rate of increase of A is directly proportional to A.

(a) Write down a differential equation that is satisfied by A.

(b) The area of the pond weed is initially 4 m² and one month later the area covered is 5 m².
Find an expression for A in terms of t.

· ·

Answer

1 (a) $\dfrac{dA}{dt} = kA$

$k > 0$
since A increases with time

(b) $\dfrac{dA}{dt} = kA$

Separating the variables and integrating gives:

$$\int \frac{1}{A} \, dA = k \int dt$$

and $\ln A = kt + C$ \hfill (1)

When $t = 0$, $A = 4$ and when $t = 1$, $A = 5$

Substituting these values into (1), we obtain:

$\ln 4 = C$ \hfill (2)

$\ln 5 = k + C$ \hfill (3)

Then $C = \ln 4$

and $k = \ln 5 - \ln 4 = \ln \dfrac{5}{4} = 0.2231$

Substituting for k and C in (1) gives:

$$\ln A = 0.2231t + \ln 4$$

$\therefore$ $\ln A - \ln 4 = 0.2231t$

$$\ln \frac{A}{4} = 0.2231t$$

Use one of the laws of logarithms:

$\ln a - \ln b = \ln \left(\dfrac{a}{b}\right)$

Taking exponentials of both sides

$$\frac{A}{4} = e^{0.2231t}$$

and $A = 4e^{0.2231t}$

2   The value of an electronic component may be modelled as a continuous variable. The value of the component at time t years is £P. The rate of decrease of P is directly proportional to P^3.

(a) Write down a differential equation that is satisfied by P.

(b) The value of the component when $t = 0$ is £20. Show that

$$\frac{1}{P^2} = \frac{1}{400} + At$$

where A is a positive constant.

(c) Given that the value of the component when $t = 1$ is £10, find the time when the value is £5.

Answer

The negative sign is introduced because P decreases with time. The constant k is then positive.

Express $\frac{1}{P^3}$ in index form so it can be integrated easily.

2   (a) $\dfrac{dP}{dt} = -kP^3$

(b) Separating variables and integrating gives:

$$\int \frac{1}{P^3}\, dP = -k \int dt$$

$$\int P^{-3}\, dP = -k \int dt$$

$$-\frac{1}{2P^2} = -kt + c$$

When $t = 0$, $P = £20$

The two known values are entered into the equation so that the value of the constant C can be found.

$$-\frac{1}{2(20)^2} = 0 + c, \text{ so } c = -\frac{1}{800}$$

Hence $-\dfrac{1}{2P^2} = -kt - \dfrac{1}{800}$

Multiplying through by -2 we obtain:

$$\frac{1}{P^2} = \frac{1}{400} + 2kt$$

A is >0 because $k > 0$.

Letting $A = 2k$ we obtain:

$$\frac{1}{P^2} = \frac{1}{400} + At$$

(c) When $t = 1$, $P = 10$

So $\dfrac{1}{100} = \dfrac{1}{400} + A$ giving $A = \dfrac{3}{400}$

Hence $\dfrac{1}{P^2} = \dfrac{1}{400} + \dfrac{3}{400}t$

When $P = 5$, $\dfrac{1}{25} = \dfrac{1}{400} + \dfrac{3}{400}t$ giving $t = 5$

3 The value, £V, of a car may be modelled as a continuous variable.
At time t years, the rate of decrease of V is directly proportional to V^2.

(a) Write down a differential equation satisfied by V.

(b) Given that $V = 12\,000$ when $t = 0$, show that

$$V = \frac{12\,000}{at + 1}$$

where a is a constant.

(c) The value of the car at the end of two years is £9000. Find the value of the car at the end of four years.

· ·

Answer

3 (a) $\dfrac{\mathrm{d}V}{\mathrm{d}t} = -kV^2$

(b)
$$\int \frac{1}{V^2}\,\mathrm{d}V = -k\int \mathrm{d}t$$

$$\int V^{-2}\,\mathrm{d}V = -k\int \mathrm{d}t$$

$$-\frac{1}{V} = -kt + c$$

Separating the variables and integrating.

When $t = 0$, $V = 12\,000$

$$-\frac{1}{12\,000} = c$$

The value for c is substituted back into the original equation.

$$-\frac{1}{V} = -kt - \frac{1}{12\,000}$$

Multiplying both sides by $-12\,000V$ gives:

$$12\,000 = 12\,000Vkt + V$$

$$12\,000 = V(12\,000kt + 1)$$

$$V = \frac{12\,000}{12\,000kt + 1}$$

Let $a = 12\,000k$

Hence $\qquad V = \dfrac{12\,000}{at + 1}$

(c) $V = 9000$ when $t = 2$

$$9000 = \frac{12\,000}{2a + 1} \quad \text{so} \quad 2a + 1 = \frac{12\,000}{9000} \quad \text{giving} \quad a = \frac{1}{6}$$

Hence $\qquad V = \dfrac{12\,000}{\frac{1}{6}t + 1}$

When $t = 4$, $\qquad V = \dfrac{12\,000}{\frac{1}{6}(4) + 1} = £7200$

Test yourself

1 The value of a car decreases with time. When the car has a value of £V after t months, the value decreases at a rate which is proportional to V.
(a) Write down a differential equation relating V and t.   [2]
(b) If the car has an initial value of £10 000, solve the differential equation and show that

$$V = 10\,000e^{-kt} \text{ where } k \text{ is a positive constant.}$$ [3]

(c) The value of the car is expected to be £4000 after 48 months.
Calculate:
(i)   the value to the nearest pound of the car when it is 12 months old,
(ii) the age of the car, to the nearest month, when its value is £3000.   [4]

2 A lawn contains some clover. The area covered by the clover at time t years is $C\,\text{m}^2$. The rate of increase of C is directly proportional to C.
(a) Write down a differential equation that is satisfied by C.   [2]

(b) The area of the lawn initially covered by the clover is $0.90\,\text{m}^2$ and two years later the area covered is $8\,\text{m}^2$. Find an expression for C in terms of t   [5].

3 An object of mass 50 kg moves in a straight horizontal line under the action of a constant horizontal force of magnitude 1600 N acting along the line. The resistance to motion of the object is proportional to time t seconds. At time t seconds, the velocity of the object is $v\,\text{m s}^{-1}$ and at time $t = 2$, it is moving with velocity $41\,\text{m s}^{-1}$ and acceleration $-4\,\text{m s}^{-2}$.
(a) Show that v satisfies the differential equation $\dfrac{\mathrm{d}v}{\mathrm{d}t} = 32 - 18t$.   [4]

(b) Find an expression for v in terms of t and determine the times when the velocity of the object is $28\,\text{m s}^{-1}$.   [4]

4 The rate of change of a population of a colony of bacteria is proportional to the size of the population P, with constant of proportionality k. At time $t = 0$ (hours), the size of the population is 10.
(a) Find an expression, in terms of k, for P at time t.   [6]

(b) Given that the population doubles after 1 hour, find the time required for the population to reach 1 million.   [3]

Summary

Check you know the following facts:

Formation of simple differential equations

If a quantity x has a rate of **increase** in x that is proportional to x, then this can be written as a differential equation by including a constant of proportionality k as

$$\frac{dx}{dt} = kx \qquad k > 0$$

If a quantity x has a rate of **decrease** in x that is proportional to x, then this can be written as a differential equation by including a constant of proportionality k as

$$\frac{dx}{dt} = -kx \qquad k > 0$$

Separating variables and integrating

If $\quad \dfrac{dm}{dt} = -km \quad$ and $\quad m = m_0$ at $t = 0$,

then variables can be separated and integrated as follows:

$$\int \frac{1}{m}\, dm = -k \int dt$$

$$\therefore \ln m = -kt + c \tag{1}$$

When $t = 0$, $\ m = m_0$

Substituting these values in (1), we obtain

$$c = \ln m_0$$

Hence we obtain:

$$\ln m = -kt + \ln m_0$$

$$\ln m - \ln m_0 = -kt$$

$$\ln\left(\frac{m}{m_0}\right) = -kt$$

Taking exponentials of both sides

> This is done to remove the ln from the left-hand side.

$$\frac{m}{m_0} = e^{-kt}$$

Hence $\qquad m = m_0 e^{-kt}$

Similarly, if

$$\frac{dP}{dt} \propto f(P)$$

> Usually, $f(P) = P^n$ where n is a constant.

$$\frac{dP}{dt} = kf(P)$$

> $k > 0$ if P increases with time
> and
> $k < 0$ if P decreases with time.

Then $\quad \displaystyle\int \frac{1}{f(P)}\, dP = \int k\, dt$

Test yourself answers

Topic 1

1 (a) $A' \cap B$
   (b) B'
   (c) $(A \cup B)'$
   (d) A'

You know they are male so the denominator is the number of males in the table (i.e. 31). The numerator is the number of males out of the 31 who prefer Indian food (i.e. 17)

2 (a) P(Prefer Indian food) $= \dfrac{29}{50}$
   (b) P(Female and prefer Chinese food) $= \dfrac{7}{50}$
   (c) P(Prefer Indian food, given that they are male) $= \dfrac{17}{31}$

3 (a)

	R	R'	**Totals**
P	6	6	**12**
P'	20	20	**40**
Totals	**26**	**26**	**52**

   (b) (i) $P(P \cap R)$ means the probability of the card being a red picture card. Look for the intersection of the P row and the R column. There are 6 such cards.
   Hence, $P(P \cap R) = \dfrac{6}{52} = \dfrac{3}{26}$

   (ii) $P(P' \cap R')$ is found by looking at the intersection of the row for P' with the column for R' and then dividing this by the total number of cards in the pack (i.e. 52).
   Hence, $P(P' \cap R') = \dfrac{20}{52} = \dfrac{5}{13}$

   (iii) $P(P|R)$ is the probability of obtaining a picture card given that a red card has been chosen. This means we only consider choosing out of half the pack (as half the pack are red).
   If we know that a red card has been chosen, there are 6 red picture cards out of 26 red cards.
   Hence, $P(P|R) = \dfrac{6}{26} = \dfrac{3}{13}$

   (iv) $P(P'|R)$ means not choosing a picture card given that a red card has been chosen. There are 20 red cards that are not picture cards. You can see this from the table.
   Hence, $P(P'|R) = \dfrac{20}{26} = \dfrac{10}{13}$

4 (a) If events are independent, $P(A \cap B) = P(A)\,P(B) = 0.6 \times 0.5 = 0.3$
   Now, $P(A \cup B) = P(A) + P(B) - P(A \cap B)$
        $0.9 = 0.6 + 0.5 - P(A \cap B)$
     $P(A \cap B) = 0.2$
   Now $0.3 \neq 0.2$, so the events are not independent.

(b) (i) $P(A') = 0.3 + 0.1 = 0.4$

(ii) $P(A \cup B)' = 0.1$

(iii) $P(A' \cap B) = 0.3$

(iv) $P(A|B) = \dfrac{0.2}{0.2 + 0.3} = 0.4$

(v) $P(A'|B) = \dfrac{0.3}{0.2 + 0.3} = \dfrac{0.3}{0.5} = 0.6$

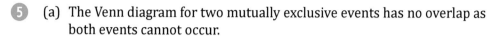

You could also use the fact that

$P(A'|B) = 1 - P(A|B)$
$\qquad\quad = 1 - 0.4 = 0.6$

5 (a) The Venn diagram for two mutually exclusive events has no overlap as both events cannot occur.

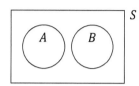

$$P(A \cup B) = P(A) + P(B) = 0.3 + 0.4 = 0.7$$

(b) $P(A \cap B) = 0.3 \times 0.4 = 0.12$

$P(A \cup B) = P(A) + P(B) - P(A \cap B)$
$\qquad\qquad = 0.3 + 0.4 - 0.12$
$\qquad\qquad = 0.58$

(c) $P(A \cap B) = P(B)\,P(A|B) = 0.4 \times 0.25 = 0.1$

$P(A \cup B) = P(A) + P(B) - P(A \cap B)$
$\qquad\qquad = 0.3 + 0.4 - 0.1$
$\qquad\qquad = 0.6$

6 (a) Now $P(A \cap B) = P(A) + P(B) - P(A \cup B)$
$\qquad\qquad\qquad = 0.4 + 0.5 - P(A \cup B)$
$\qquad\qquad\qquad = 0.9 - 2P(A \cap B)$

Hence $3P(A \cap B) = 0.9$
$\qquad\quad P(A \cap B) = 0.3$

(b) $P(A|B) = \dfrac{P(A \cap B)}{P(B)}$

$\qquad\qquad = \dfrac{0.3}{0.5} = \dfrac{3}{5}$

$\qquad\qquad = 0.6$

(c)

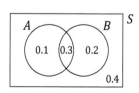

$P(B|A') = \dfrac{P(B \cap A')}{P(A')}$

$\qquad\qquad = \dfrac{0.2}{0.2 + 0.4} = \dfrac{0.2}{0.6}$

$\qquad\qquad = \dfrac{1}{3}$

7 (a)

	P	P'	**Totals**
C	21	49	**70**
C'	40	10	**50**
Totals	**61**	**59**	**120**

(b)

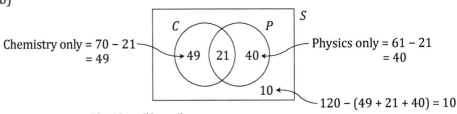

Chemistry only = 70 − 21 = 49

Physics only = 61 − 21 = 40

120 − (49 + 21 + 40) = 10

(c) (i) $P(C') = \dfrac{40+10}{120} = \dfrac{50}{120} = \dfrac{5}{12}$

(ii) $P(C' \cap P') = \dfrac{10}{120} = \dfrac{1}{12}$

(iii) $P(P|C) = \dfrac{21}{49+21} = \dfrac{21}{70} = \dfrac{3}{10}$

(iv) $P(C'|P) = \dfrac{40}{21+40} = \dfrac{40}{61}$

8 (a) If events are independent, $P(A \cap B) = P(A)\,P(B) = 0.3 \times 0.5 = 0.15$
Now, $\quad P(A \cup B) = P(A) + P(B) - P(A \cap B)$
$\quad\quad\quad 0.7 = 0.3 + 0.5 - P(A \cap B)$
$\quad\quad\quad P(A \cap B) = 0.1$
Now $0.15 \neq 0.1$, so the events are not independent.

(b) (i) $P(A \cap B) = P(B)\,P(A|B)$

Hence, $\quad P(A|B) = \dfrac{P(A \cap B)}{P(B)}$

$\quad\quad\quad = \dfrac{0.1}{0.5}$

$\quad\quad\quad = \dfrac{1}{5}$

Now $\quad P(A'|B) = 1 - P(A|B)$

$\quad\quad\quad = 1 - \dfrac{1}{5}$

$\quad\quad\quad = \dfrac{4}{5}$

(ii) $P(A \cup B) = P(A) + P(B) - P(A \cap B)$
So, replacing A with A' we obtain
$\quad P(A' \cup B) = P(A') + P(B) - P(A' \cap B)$
$\quad\quad\quad = 0.7 + 0.5 - (P(B) - P(A \cap B))$
$\quad\quad\quad = 1.2 - (0.5 - 0.1)$
$\quad\quad\quad = 0.8$

You could use the following alternative Venn diagram method for part (b)(i) and (ii)

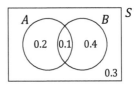

(b) (i) $P(A'|B) = \dfrac{0.4}{0.1 + 0.4}$

$\qquad\qquad = \dfrac{0.4}{0.5}$

$\qquad\qquad = \dfrac{4}{5}$

> This is everything outside A joined to everything in B.

(ii) $P(A' \cup B) = 0.4 + 0.3 + 0.1$

$\qquad\qquad\quad = 0.8$

Topic 2

1 $P(L \le 4) = P(A \le 16)$

$P(A \le 16) = \dfrac{d - c}{b - a} = \dfrac{16 - 15}{20 - 15}$

$\qquad\qquad\quad = 0.2$

> The square has an area ≤ 16 if the side $L \le 4$.

2 (a) The following formula is used to find the standardised variable z:

$$z = \frac{x - \mu}{\sigma}$$

$$z = \frac{10.5 - 10}{2}$$

$$= 0.25$$

So $\qquad P(X \le 10.5) = P(Z \le 0.25)$

$\qquad\qquad\qquad\qquad = 0.59871$

(b) $P(X \ge x) = 0.1$, so $P(X < x) = 0.9$

$\qquad\qquad\qquad P(Z < 1.28) \approx 0.9$

$$z = \frac{x - \mu}{\sigma}$$

$$1.28 = \frac{x - 10}{2}$$

$$2.56 = x - 10$$

$$x = 12.56$$

> Note that $P(X \ge x) = 0.1$ is in the upper tail.

3 (a) $E(X) = \frac{1}{2}(a + b)$

$\qquad\qquad = \frac{1}{2}(1.5 + 7.5)$

$\qquad\qquad = 4.5$

$\quad Var(X) = \frac{1}{12}(b - a)^2$

$\qquad\qquad = \frac{1}{12}(7.5 - 1.5)^2$

$\qquad\qquad = 3$

$\quad$ Standard deviation $\sigma = \sqrt{3}$

$\qquad\qquad\qquad\qquad = 1.732$

> The standard deviation is the square root of the variance.

(b)

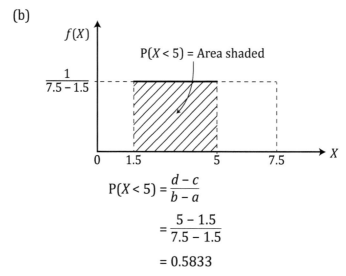

$$P(X < 5) = \frac{d - c}{b - a}$$

$$= \frac{5 - 1.5}{7.5 - 1.5}$$

$$= 0.5833$$

④ Note that $Z \sim N(0, 1)$ is the standard normal distribution and we do not need to calculate the z-values, as they are given.
We draw the first graph, shading the area between the two z-values. This area represents the required probability.

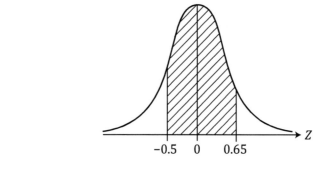

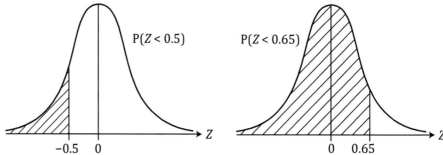

The two graphs next to each other show that we can work out the required area as follows:

$$P(-0.5 < z < 0.65) = P(z < 0.65) - P(z < -0.5)$$

Now $\quad\quad P(z < -0.5) = 1 - P(z < 0.5)$

Hence, $\quad P(-0.5 < z < 0.65) = P(z < 0.65) - (1 - P(z < 0.5))$

$$= 0.74215 - (1 - 0.69146)$$

$$= 0.43361$$

5 (a) $z = \dfrac{x - \mu}{\sigma}$

$\quad = \dfrac{130 - 120}{25}$

$\quad = 0.4$

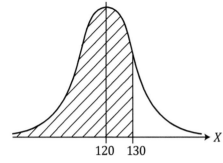 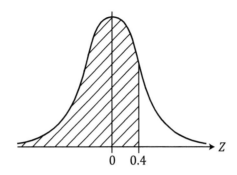

$$P(X < 130) = P(Z < 0.4)$$
$$= 0.65542$$

(b) $z = \dfrac{x - \mu}{\sigma}$

$\quad = \dfrac{100 - 120}{25}$

$\quad = -0.8$

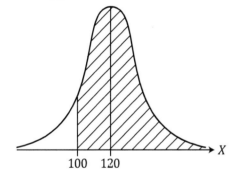 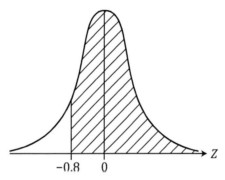

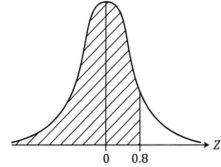

$$P(X > 100) = P(Z > -0.8)$$
$$= P(Z < 0.8)$$
$$= 0.78814$$

(c) For $x = 90$, $z = \dfrac{x - \mu}{\sigma} = \dfrac{90 - 120}{25} = -1.2$

For $x = 130$, $z = \dfrac{x - \mu}{\sigma} = \dfrac{130 - 120}{25} = 0.4$

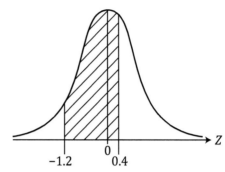

Note that P($z < 0.4$) was found in part (a).

$$P(-1.2 < z < 0.4) = P(z < 0.4) - (1 - P(z < 1.2))$$
$$= 0.65542 - (1 - 0.88493)$$
$$= 0.54035$$

6 (a) For $x = 227$, $z = \dfrac{x - \mu}{\sigma} = \dfrac{227 - 240}{15} = -0.8667$

Note that using the tables the nearest value to 0.8667 is 0.87 so we use this value when finding the probability.

$$P(z < -0.8667) = P(z > 0.8667)$$
$$= 1 - P(z < 0.8667)$$
$$= 1 - 0.80785$$
$$= 0.19215$$

(b) For $x = 227$, $z = \dfrac{x - \mu}{\sigma} = \dfrac{227 - 240}{15} = -0.8667 = -0.87$

For $x = 240$, $z = \dfrac{x - \mu}{\sigma} = \dfrac{240 - 240}{25} = 0$

$$P(227 < X < 240) = P(-0.87 < Z < 0)$$

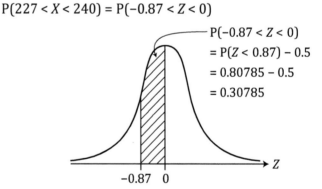

P($-0.87 < Z < 0$)
= P($Z < 0.87$) – 0.5
= 0.80785 – 0.5
= 0.30785

Hence P($227 < X < 240$) = 0.30785

7 Need to find P($X > 90$)

For $x = 90$, $z = \dfrac{x - \mu}{\sigma} = \dfrac{90 - 65}{15} = 1.67$

$$P(X > 90) = P(Z > 1.67) = 1 - P(Z < 1.67) = 1 - 0.95254 = 0.04766$$

8 $P(X < 170) = 0.14$ and as the probability is less than 0.5, the z-value must be on the left of the mean (i.e. the z-value must be negative).

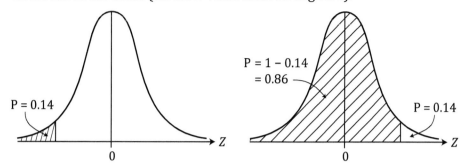

We need to find the z-value that gives a probability of 0.86. To find this we look at the probabilities in the body of the table nearest to 0.86 and then read off the corresponding z-value.

The nearest probability to 0.86 in the table is 0.85993, which gives a z-value of 1.08, but as it is on the left of the mean the z-value we must use is -1.08.

$$z = \frac{x - \mu}{\sigma}$$

$$-1.08 = \frac{170 - \mu}{\sigma}$$

$$-1.08\sigma = 170 - \mu \qquad (1)$$

$P(X > 200) = 0.03$, so this must correspond to a z-value to the right of the mean.

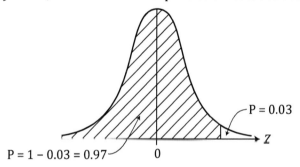

The probability we need to look up in the table is $1 - 0.03 = 0.97$
Looking up the probability of 0.97 from the tables we find that the z-value is 1.88.

$$z = \frac{x - \mu}{\sigma}$$

$$1.88 = \frac{200 - \mu}{\sigma}$$

$$1.88\sigma = 200 - \mu \qquad (2)$$

Solving these two equations simultaneously:
Subtracting equation (1) from equation (2) we obtain

$$2.96\sigma = 30$$

Hence $\qquad\qquad \sigma = 10.14$ minutes
So $\qquad\qquad 1.88 \times 10.14 = 200 - \mu$
Hence $\qquad\qquad \mu = 181$ minutes

⑨ We need to find the probability $P(35 < X < 45)$

For $X = 35$, $z = \dfrac{x - \mu}{\sigma} = \dfrac{35 - 30}{5} = 1$

For $X = 45$, $z = \dfrac{x - \mu}{\sigma} = \dfrac{45 - 30}{5} = 3$

Hence, $P(35 < X < 45) = P(1 < Z < 3)$

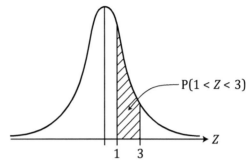

From the graph we can see $P(1 < Z < 3) = P(Z < 3) - P(Z < 1)$
Now using the table, $P(Z < 3) = 0.99865$ and $P(Z < 1) = 0.84134$

$$P(1 < Z < 3) = P(Z < 3) - P(Z < 1)$$
$$= 0.99865 - 0.84134$$
$$= 0.15731$$

The probability that the length of time between charges is between 35 and 45 hours = 0.15731

⑩ (a)

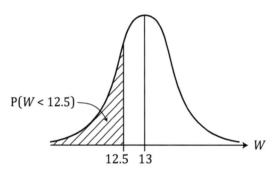

The calculator is set to 'Normal CD' and the following parameters entered:
Lower = 1×10^{-99}
Upper = 12.5
$\sigma = 0.8$
$\mu = 13$

Using the calculator, $P(W < 12.5) = 0.2660$

(b) $P(\text{Two bags} < 12.5\text{ kg}) = 0.2660 \times 0.2660$
$= 0.071$ (2 s.f.)

Topic 3

1   Notice that in this question we are looking for any correlation (i.e. positive or negative).

Null hypothesis is $\mathbf{H_0} : r = 0$

Alternative hypothesis is $\mathbf{H_1} : r \neq 0$

This is a two-tailed test so looking up 5% significance level for a two-tailed test and $n = 15$, the critical value is read off from the 'Critical values of the product moment correlation coefficient' table.

From the table, the critical value is 0.5140

The correlation coefficient of 0.55 > the critical value of 0.5140.

The result is significant at the 5% level of significance. There is evidence to reject the null hypothesis in favour of the alternative hypothesis indicating evidence of correlation.

Hence, there is evidence that the number of shoppers per week is correlated to the number of special offers.

> We have used the letter r as the correlation coefficient as it is based on a sample and not the population.

2   Hence we can write, $\mathbf{H_0} : \mu = 60$

$\mathbf{H_1} : \mu > 60$

Need to calculate the p-value

> The values for $\overline{X}$, μ and n are entered into
> $$z = \frac{\overline{X} - \mu}{\frac{\sigma}{\sqrt{n}}}$$

p-value $= P(\overline{X} > 61)$ assuming the null hypothesis

$$= P\left(z > \frac{61 - 60}{\frac{3}{\sqrt{36}}}\right)$$

$$= P(z > 2)$$

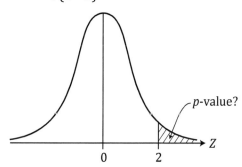

p-value $= 1 - P(z < 2)$

$$= 1 - 0.97725$$

$$= 0.02275$$

> This p-value is compared with the following:
>
> $p < 0.01$; there is very strong evidence for rejecting $\mathbf{H_0}$
>
> $0.01 \leq p \leq 0.05$; there is strong evidence for rejecting $\mathbf{H_0}$
>
> $p > 0.05$; there is insufficient evidence for rejecting $\mathbf{H_0}$.

Now the p-value of 0.02275 is in the range $0.01 \leq p \leq 0.05$ so there is strong evidence for rejecting $\mathbf{H_0}$. This means that there is evidence that the mean is greater than 60.

The values for $\overline{X}$, μ and n are entered into

$$z = \frac{\overline{X} - \mu}{\frac{\sigma}{\sqrt{n}}}$$

3 The null hypothesis is that the mean lifespan is 10 years.
The alternative hypothesis is that the mean lifespan is greater than 10 years.
Hence we can write, $\mathbf{H_0} : \mu = 10$
$\qquad\qquad\qquad \mathbf{H_1} : \mu > 10$
Need to calculate the p-value

$$p\text{-value} = P(\overline{X} > 11) \quad \text{assuming the null hypothesis}$$

$$= P\left(z > \frac{11 - 10}{\frac{2}{\sqrt{25}}}\right)$$

$$= P(z > 2.5)$$

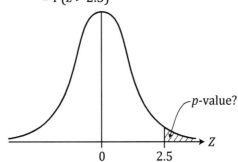

p-value?

$$p\text{-value} = 1 - P(z < 2.5)$$
$$= 1 - 0.99379$$
$$= 0.00621$$

As this p-value < 0.01, there is very strong evidence to reject $\mathbf{H_0}$.
Hence there is very strong evidence that the mean lifespan is greater than 10 years.

Notice that the mean μ is not given in the question. However, we know from the information about the alternative hypothesis that $\mu \neq 12$, so μ must be 12 kg.

The values for $\overline{X}$, μ and n are entered into

$$z = \frac{\overline{X} - \mu}{\frac{\sigma}{\sqrt{n}}}$$

4 The null hypothesis is that the mean weight = 12 kg
The alternative hypothesis is that the mean weight is not equal to 12 kg.
Hence we can write, $\mathbf{H_0} : \mu = 12$
$\qquad\qquad\qquad \mathbf{H_1} : \mu \neq 12$
Need to calculate the p-value

$$p\text{-value} = P(\overline{X} < 11.8) \quad \text{assuming the null hypothesis}$$

$$= P\left(z > \frac{11.8 - 12}{\frac{1.8}{\sqrt{20}}}\right)$$

$$= P(z < -0.4969)$$
$$= P(z < -0.50) \text{ (2 d.p.)}$$
$$P(Z < -0.50) = 1 - P(Z < 0.5)$$
$$= 1 - 0.69146$$
$$= 0.30854$$

As this is a two-tailed test we need to double this probability, so
$$p\text{-value} = 2 \times 0.30854$$
$$= 0.61708$$

As $p > 0.05$; there is insufficient evidence for rejecting $\mathbf{H_0}$.
Hence there is evidence that the machine is filling sacks with the correct weight and therefore does not need adjusting.

5 (a) The two hypotheses used are:
$$\mathbf{H_0} : \mu = 15\,000$$
$$\mathbf{H_1} : \mu < 15\,000$$

Let $\overline{X}$ = mean lifetime of the sampled bulbs
The mean value of the sample assumed to be normally distributed, so we have:
$$\overline{X} \sim N\left(14\,500, \frac{2500^2}{100}\right)$$
Now we standardise the normal distribution by finding the z-value.
Now
$$z = \frac{\overline{X} - \mu}{\frac{\sigma}{\sqrt{n}}}$$
$$= \frac{14\,500 - 15\,000}{\frac{2500}{\sqrt{100}}}$$
$$= -2$$
$$P(Z < -2) = 1 - P(Z < 2)$$
$$= 1 - 0.97725$$
$$= 0.0228 \text{ or } 2.28\%$$

(b) We first find the z-value that gives a significance of 1% (i.e. 0.01).

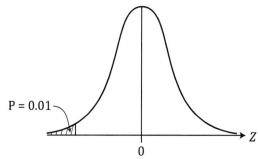

P = 0.01

The value of z for which $P(Z < z) = 0.01$ is -2.326
We can now use the formula
$$z = \frac{\overline{X} - \mu}{\frac{\sigma}{\sqrt{n}}}$$
to find the critical value for the mean lifetime for the bulb.
$$-2.326 = \frac{\overline{X} - 15\,000}{\frac{2500}{\sqrt{100}}}$$
$$\overline{X} = 14\,418.8 \text{ hours}$$
Critical value for lifetime of bulbs = 14 418.8 hours

6 The null hypothesis is that the mean weight is 20.5 kg
The alternative hypothesis is that the mean weight is less than 20.5 kg
Hence we can write
$$\mathbf{H_0} : \mu = 20.5$$
$$\mathbf{H_1} : \mu < 20.5$$

Let X be the mean weight of a suitcase which is normally distributed.

Assuming $\mathbf{H_0}$ if $X \sim N(\mu, \sigma^2)$ then the sample weight mean, $\overline{X}$, is normally

distributed so we can say $\overline{X} \sim N\left(\mu, \frac{\sigma^2}{n}\right)$.

Now we standardise the normal distribution by finding the z-value.

Now
$$z = \frac{\bar{x} - \mu}{\frac{\sigma}{\sqrt{n}}}$$

$$= \frac{19.6 - 20.5}{\frac{\sqrt{0.8}}{\sqrt{10}}}$$

$$= -3.1820$$

$$P(Z < -3.1820) = 1 - P(Z < 3.1820)$$
$$= 1 - 0.99926$$
$$= 0.00074$$

As this p-value of $0.00074 < 0.01$, there is very strong evidence to reject $\mathbf{H_0}$. Hence there is very strong evidence that the mean weight of suitcases has decreased.

7 (a) $\mathbf{H_0} : \mu = 8$
$\mathbf{H_1} : \mu > 8$

(b) Let the breaking strain = X kg.

If X is normally distributed then we can say that the sample mean, $\bar{X}$, is normally distributed.

Using $Z = \frac{\bar{x} - \mu}{\frac{\sigma}{\sqrt{n}}}$ to calculate the z-value, we have:

$$z\text{-value} = \frac{8.2 - 8}{\frac{0.9}{\sqrt{30}}} = 1.22$$

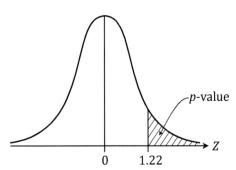

$$P(Z \geq 1.22) = 1 - P(Z \leq 1.22)$$
$$= 1 - 0.88877$$
$$= 0.11123$$

The p-value of 0.11123 is greater than 0.05 which means there is little evidence to reject the null hypothesis meaning there is insufficient evidence to conclude that the breaking strain has increased due to the new process.

8 (a) Null hypothesis $\mathrm{H_0} : \rho = 0$
Alternative hypothesis is $H_1 : \rho \neq 0$

(b) PMCC = 0.75 and $n = 40$.
Because we are looking at $H_1 : \rho \neq 0$ we need to consider $H_1 : \rho < 0$ and $H_1 : \rho > 0$ so this is a two-tailed test.
We now find the critical value of the product moment correlation coefficient by looking it up in Table 9 of the statistical tables.

We are conducting a two-tailed test at a 5% significance level, so we look along the column to the 5% two tail test. We then look for a value of n equal to 40 (i.e. the number of pairs of values) and then look for the intersection of the column and row. The critical value is ±0.3120. (Note that we include the ± as we are looking at both tails of the normal distribution curve as this is a two-tailed test.)

The test statistic the PMCC, 0.75 > the critical value, 0.3120.

This result is significant and this means the null hypothesis is rejected and that the product moment correlation coefficient is greater than 0 and that there is evidence at the 5% level of significance that there is correlation between the two variables.

Topic 4

1 (a) $v = \dfrac{ds}{dt}$

$= 36t^2$

(b) $a = \dfrac{dv}{dt}$

$= 72t$

When $t = 2$, $a = 72 \times 2 = 144$ m

> $t = 2$ is substituted into the expression for a.

2 (a) $v = \int a \, dt$

$= \int (3 - 0.1t) \, dt$

$= 3t - \dfrac{0.1t^2}{2} + c$

When $t = 0$, $v = 0$ so $0 = 3(0) - \dfrac{0.1(0)^2}{2} + c$.

Solving gives $c = 0$.

The expression for the velocity is $v = 3t - \dfrac{0.1t^2}{2}$.

(b) When $t = 10$, $v = 3(10) - \dfrac{0.1(10)^2}{2}$

$v = 30 - 5 = 25$ m s^{-1}

(c) When $t = 30$, $a = 3 - 0.1(30) = 0$

As the acceleration is zero it will travel at constant speed/velocity.

(d) $s = \int v \, dt$

$= \int \left(3t - \dfrac{0.1t^2}{2}\right) dt$

$= \dfrac{3t^2}{2} - \dfrac{0.1t^3}{6} + c$

When $t = 0$, $s = 0$, hence $c = 0$.

so substituting these values into the above expression gives

$s = \dfrac{3t^2}{2} - \dfrac{0.1t^3}{6}$

When $t = 30$, $s = \dfrac{3(30)^2}{2} - \dfrac{0.1(30)^3}{6}$

$= 1350 - 450 = 900$ m

3 $s = \int v \, dt$

$= \int (6t^2 + 4) \, dt$

$= \dfrac{6t^3}{3} + 4t + c$

$= 2t^3 + 4t + c$

When $t = 0$, $s = 0$ so we have $0 = 2(0)^3 + 4(0) + c$ and solving gives $c = 0$.

Hence $s = 2t^3 + 4t$

When $t = 2$ s, $s = 2(2)^3 + 4(2) = 24$ m

When $t = 5$ s, $s = 2(5)^3 + 4(5) = 270$ m

Distance travelled between the times $t = 2$ s and $t = 5$ s is $270 - 24 = 246$ m

4   (a) (i) $a = \dfrac{dv}{dt}$

$= 12t - 2$

(ii) When $t = 1$, $a = 12(1) - 2 = 10 \text{ m s}^{-2}$

> The particle is at the origin at $t = 0$ so we know $s = 0$ as the origin is the point from which the displacement is measured.

(b) $s = \int v \, dt$

$= \int (6t^2 - 2t + 8) \, dt$

$= \dfrac{6t^3}{3} - \dfrac{2t^2}{2} + 8t + c$

$= 2t^3 - t^2 + 8t + c$

When $t = 0$, $s = 0$ so $0 = 2(0)^3 - (0)^2 + 8(0) + c$

Solving gives $c = 0$.

Hence, the expression for the displacement is

$s = 2t^3 - t^2 + 8t$

5   (a) $v = 2t(t - 6)$

When $v = 0$, $t = 0$ or 6

We need to find the values of t for which $2t(t - 6) < 0$

The curve for $2t(t - 6)$ will be $\cup$-shaped and will intersect the x-axis at 0 and 6.

We want the values of t for which the curve will be below the x-axis.

Hence we have $0 < t < 6$.

> Always be guided by what you did in the previous part of the question to what you have to do next. Here we know the velocity is negative between the times 0 and 6 s. If there is a negative part to the velocity–time graph, the area of the region will be negative, so the displacement will be negative. We need to change this into a positive value as we want the distance.

(b) $s = \displaystyle\int_{6}^{9} v \, dt = \int_{6}^{9} (2t^2 - 12t) \, dt$

$= \left[\dfrac{2t^3}{3} - \dfrac{12t^2}{2} \right]_{6}^{9}$

$= \left[\dfrac{2t^3}{3} - 6t^2 \right]_{6}^{9}$

$= \left[\left(\dfrac{2 \times 729}{3} - 486 \right) - \left(\dfrac{2 \times 216}{3} - 216 \right) \right]$

$= 72$

> The minus sign is placed in front of the integral because we know the displacement is negative but the distance should be positive as it is a scalar quantity.

$s = \displaystyle\int_{0}^{6} v \, dt = -\left[\dfrac{2t^3}{3} - 6t^2 \right]_{0}^{6}$

$= 72$

Total distance = $72 + 72 = 144$

Topic 5

1 $\mathbf{a} = \dfrac{\mathbf{F}}{m} = \dfrac{(4t-3)\mathbf{i} + (3t^2 - 5t)\mathbf{j}}{0.5} = (8t-6)\mathbf{i} + (6t^2 - 10t)\mathbf{j}$

$$\mathbf{v} = \int \mathbf{a}\,\mathrm{d}t$$

$$= \int \left[(8t-6)\mathbf{i} + (6t^2 - 10t)\mathbf{j}\right]\mathrm{d}t$$

$$= (4t^2 - 6t)\mathbf{i} + (2t^3 - 5t^2)\mathbf{j} + \mathbf{c}$$

When $t = 0$, $\qquad \mathbf{v} = 8\mathbf{i} - 7\mathbf{j}$

Hence $\qquad 8\mathbf{i} - 7\mathbf{j} = \mathbf{c}$

So $\qquad \mathbf{v} = (4t^2 - 6t)\mathbf{i} + (2t^3 - 5t^2)\mathbf{j} + 8\mathbf{i} - 7\mathbf{j}$

$$= (4t^2 - 6t + 8)\mathbf{i} + (2t^3 - 5t^2 - 7)\mathbf{j}$$

Note that the constant of integration, **c**, is a vector.

The constant of integration, **c** is substituted back into the equation for **v**.

2 $s = \int v\,\mathrm{d}t$

$$= \int_0^{\frac{\pi}{6}} (4\cos 2t)\mathrm{d}t$$

$$= \left[2\sin 2t\right]_0^{\frac{\pi}{6}}$$

$$= 2\sin\frac{\pi}{3} - 0$$

$$= \sqrt{3}$$

$$= 1.732\,\mathrm{m}$$

Note $\sin\dfrac{\pi}{3} = \dfrac{\sqrt{3}}{2}$

3 $\mathbf{v} = \dfrac{\mathrm{d}\mathbf{r}}{\mathrm{d}t} = (1 + 4t)\mathbf{i} + (3t - 2)\mathbf{j}$

$\mathbf{a} = \dfrac{\mathrm{d}\mathbf{v}}{\mathrm{d}t} = 4\mathbf{i} + 3\mathbf{j}$ which is independent of t and therefore constant.

Magnitude of acceleration $= |\mathbf{a}| = \sqrt{4^2 + 3^2}$

$$= 5\,\mathrm{m\,s^{-2}}$$

4 (a) $\mathbf{a} = \dfrac{\mathbf{F}}{m} = \dfrac{-3\mathbf{i} + 4\mathbf{j} - 5\mathbf{k}}{2} = -1.5\mathbf{i} + 2\mathbf{j} - 2.5\mathbf{k}$

Magnitude of acceleration $|\mathbf{a}| = \sqrt{(-1.5)^2 + 2^2 + (-2.5)^2}$

$$= \sqrt{12.5}$$

$$= 3.54\,\mathrm{m\,s^{-2}}$$

(b) Using $\mathbf{s} = \mathbf{u}t + \dfrac{1}{2}\mathbf{a}t^2$

$$= (3\mathbf{i} - 2\mathbf{j} + \mathbf{k})2 + \frac{1}{2} \times (-1.5\mathbf{i} + 2\mathbf{j} - 2.5\mathbf{k})4$$

$$= 6\mathbf{i} - 4\mathbf{j} + 2\mathbf{k} - 3\mathbf{i} + 4\mathbf{j} - 5\mathbf{k}$$

$$= 3\mathbf{i} - 3\mathbf{k}$$

Now at $t = 0$, the position vector is $2\mathbf{i} - 7\mathbf{j} + 9\mathbf{k}$ so this vector needs to be added to the displacement vector travelled in 2 s to find the displacement vector from the origin.

Position vector $= 3\mathbf{i} - 3\mathbf{k} + 2\mathbf{i} - 7\mathbf{j} + 9\mathbf{k}$

$$= 5\mathbf{i} - 7\mathbf{j} + 6\mathbf{k}$$

5 (a) $\mathbf{v} = \dfrac{d\mathbf{r}}{dt} = 8t\mathbf{i} - 7\mathbf{j}$

$\mathbf{a} = \dfrac{d\mathbf{v}}{dt} = 8\mathbf{i}$ which is independent of t and therefore constant.

If the acceleration is constant, then the force will be constant as

Force = mass × acceleration (i.e. $\mathbf{F} = m\mathbf{a}$).

(b) $\mathbf{F} = m\mathbf{a}$

$= 3 \times 8\mathbf{i}$

$= 24\mathbf{i}$ newtons

6 (a) $\mathbf{v} = 2\sin 3t\,\mathbf{i} + 4\cos 3t\,\mathbf{j}$

$\mathbf{a} = \dfrac{d\mathbf{v}}{dt} = 6\cos 3t\,\mathbf{i} - 12\sin 3t\,\mathbf{j}$

When $t = \dfrac{\pi}{3}$, $\quad \mathbf{a} = 6\cos 3\left(\dfrac{\pi}{3}\right)\mathbf{i} - 12\sin 3\left(\dfrac{\pi}{3}\right)\mathbf{j}$

$= 6\cos \pi\,\mathbf{i} - 12\sin \pi\,\mathbf{j}$

$= -6\mathbf{i} - 0$

$= -6\mathbf{i}\ \mathrm{m\,s}^{-2}$

(b) $\mathbf{s} = \displaystyle\int \mathbf{v}\,dt = \int \left(2\sin 3t\,\mathbf{i} + 4\cos 3t\,\mathbf{j}\right)dt$

$= -\dfrac{2}{3}\cos 3t\,\mathbf{i} + \dfrac{4}{3}\sin 3t\,\mathbf{j} + c$

When $t = 0$, $\mathbf{s} = \mathbf{i} + 2\mathbf{j}$

Hence $\quad \mathbf{i} + 2\mathbf{j} = -\dfrac{2}{3}\cos(0)\,\mathbf{i} + \dfrac{4}{3}\sin(0)\,\mathbf{j} + c$

$\mathbf{i} + 2\mathbf{j} = -\dfrac{2}{3}\mathbf{i} + 0 + c$

$c = \dfrac{5}{3}\mathbf{i} + 2\mathbf{j}$

So $\quad \mathbf{s} = -\dfrac{2}{3}\cos 3t\,\mathbf{i} + \dfrac{4}{3}\sin 3t\,\mathbf{j} + \dfrac{5}{3}\mathbf{i} + 2\mathbf{j}$

When $t = \dfrac{\pi}{3}$, $\quad \mathbf{s} = -\dfrac{2}{3}\cos 3\left(\dfrac{\pi}{3}\right)\mathbf{i} + \dfrac{4}{3}\sin 3\left(\dfrac{\pi}{3}\right)\mathbf{j} + \dfrac{5}{3}\mathbf{i} + 2\mathbf{j}$

$= \dfrac{2}{3}\mathbf{i} + 0 + \dfrac{5}{3}\mathbf{i} + 2\mathbf{j}$

$= \dfrac{7}{3}\mathbf{i} + 2\mathbf{j}$

Topic 6

1

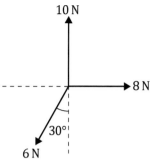

Component of the 6 N force to the left of the 8 N force = 6 sin 30°
$$= 3 \text{ N (to the left)}$$

Resolving in the direction of the 8 N force we obtain,
net resultant force $\qquad = 8 - 3$
$$= 5 \text{ N (to the right)}$$

Component of the 6 N force in the direction
opposite to the 8 N force $\qquad = 6 \cos 30°$
$$= 5.1962 \text{ N (downwards)}$$

Resolving in the direction of the 10 N force,
we obtain, net resultant force $\qquad = 10 - 5.1962$
$$= 4.8038 \text{ N (upwards)}$$

Resultant force R is found using
Pythagoras' theorem
$$R^2 = 5^2 + 4.8038^2$$
giving $\qquad R = 6.9337 \text{ N}$

$$\theta = \tan^{-1}\left(\frac{4.8038}{5}\right)$$

$$= 43.9° \text{ (to the direction of the 8 N force)}$$

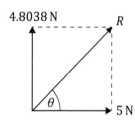

2

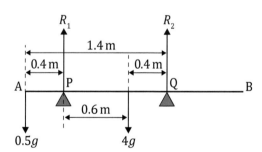

Resolving vertically, we obtain
$$R_1 + R_2 = 0.5g + 4g$$
$$R_1 + R_2 = 4.5 \times 9.8$$
$$R_1 + R_2 = 44.1 \text{ N} \qquad (1)$$
Taking moments about P, we obtain
$$4g \times 0.6 = (0.5g \times 0.4) + (R_2 \times 1)$$
Hence $\qquad R_2 = 21.56 \text{ N}$
Substituting $R_2 = 21.56$ into equation (1) we obtain
$$R_1 + 21.56 = 44.1$$
Hence $\qquad R_1 = 22.54 \text{ N}$

> Clockwise moments = anticlockwise moments.

3

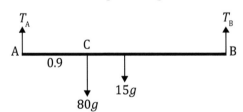

As the rod is in equilibrium, resolving vertically, we obtain
$$T_1 + T_2 = 5g + 2g$$
$$T_1 + T_2 = 7g \qquad (1)$$
Taking moments about point A we obtain
$$(5g \times 0.9) + (2g \times 1.6) = T_2 \times 1.8$$
Solving, gives $T_2 = 41.9222\,\text{N}$ $\qquad T_2 = 41.9\,\text{N}$ (correct to one decimal place)
Substituting this value for T_2 into equation (1), we obtain
$$T_1 + 41.9222 = 7 \times 9.8$$
$$T_1 = 26.6778\,\text{N} \qquad T_1 = 26.7\,\text{N} \text{ (correct to one decimal place)}$$

4 Resolving in the direction of the $Q\,\text{N}$ force, we obtain
$$Q = 9\sin 60°$$
$$= 9 \times \frac{\sqrt{3}}{2} = 7.794\,\text{N}$$
Resolving in the direction of the $P\,\text{N}$ force, we obtain
$$P + 9\cos 60° = 6$$
$$P = 6 - 9 \times 0.5 = 1.5\,\text{N}$$

5 (a) Resolving horizontally, we obtain $\qquad T_1\cos 23° = T_2\cos 40°$
$$0.9205\,T_1 = 0.7660\,T_2$$
$$T_2 = \frac{0.9205}{0.7660}\,T_1$$
$$= 1.2017\,T_1$$
Hence, $T_2 \approx 1.2\,T_1$

We can use the result that $T_2 \approx 1.2\,T_1$ to help solve the simultaneous equations.

(b) Resolving vertically, we obtain
$$T_1\sin 23° + T_2\sin 40° = 160$$
$$T_1 = \frac{160}{\sin 23° + 1.2\sin 40°}$$
$$= 137.68\,\text{N}$$

(c) The object is modelled as a particle.
The cables are modelled as light strings and therefore have no weight.

6 (a)

Taking moments about A, we obtain
$$2.8\,T_B = 80g \times 0.9 + 15g \times 1.4$$
$$T_B = 325.5\,\text{N}$$

Resolving vertically, we obtain
$$T_A + T_B = 80g + 15g$$
$$T_A = 605.5\,\text{N}$$

(b)

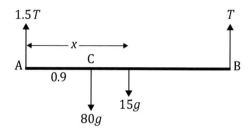

Resolving vertically, we obtain
$$1.5T + T = 95g$$
$$T = 38g$$
Taking moments about A, we obtain
$$2.8 \times T = 80g \times 0.9 + 15gx$$
$$2.8 \times 38g = 80g \times 0.9 + 15gx$$
$$x = 2.3\,\text{m}$$

7 (a)

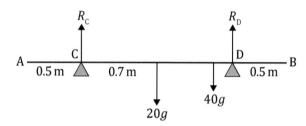

Taking moments about D, we obtain:
$$40g \times 0.1 + 20g \times 0.7 = R_C \times 1.4$$
$$R_C = 126\,\text{N}$$
Resolving vertically, we obtain
$$R_C + R_D = 40g + 20g$$
$$R_D = 462\,\text{N}$$

(b)

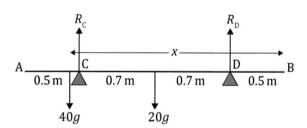

Let x be the distance from B to the person.
Taking moments about C, we obtain:
$$40g(x - 1.9) + R_D \times 1.4 = 20g \times 0.7$$
Equilibrium on point of collapse occurs when $R_D = 0$
Taking moments about C, we obtain:
$$40g(x - 1.9) = 20g \times 0.7$$
$$x = 2.25\,\text{m}$$

Note that the person must have walked past point C for the equilibrium to collapse.

Topic 7

1 (a)

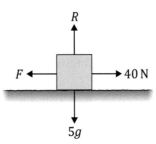

Applying Newton's second law of motion to particle P, we obtain

$$ma = T + 5g \sin 30°$$
$$5a = T + 2.5g \qquad (1)$$

Applying Newton's second law of motion to particle Q, we obtain

$$8a = 8g - T \qquad (2)$$

Adding equations (1) and (2), we obtain

$$13a = 10.5g$$
$$a = 7.9 \text{ m s}^{-2}$$

(b) Substituting $a = 7.9$ into equation (1), we obtain

$$5 \times 7.9 = T + (2.5 \times 9.8)$$
$$T = 15.1 \text{ N (correct to 1 decimal place)}$$

2 (a)

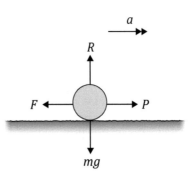

(b) Resolving vertically, we obtain $R = 5g = 5 \times 9.8 = 49$ N

(c) Applying Newton's second law to the horizontal motion, we obtain

$$ma = 40 - F$$
$$5 \times 2 = 40 - F$$

Hence $\qquad F = 30$ N

3 (a)

Resolving vertically, we have

$$R = 3g$$

$$= 3 \times 9.8$$
$$= 29.4 \text{ N}$$

$F_{MAX} = \mu R = 0.3 \times 29.4 = 8.82 \text{ N}$

F_{MAX} is the limiting friction. This is the maximum friction before movement occurs.

As the maximum frictional force (i.e. 8.82 N) is larger than the applied force P of 8 N, the frictional force F will balance P.

Hence $F = 8$ N

As P and F are balanced, there is no resultant force so the acceleration will be 0 m s^{-2} so the particle will remain at rest.

(b) $R = 29.4$ N and $F_{MAX} = 8.82$ N

P is greater than F_{MAX} so there will be a resultant force.

Resultant force $= P - F_{MAX} = 12 - 8.82 = 3.18$ N

Applying Newton's second law, we obtain

$$ma = 3.18$$
$$3a = 3.18$$

Hence $\quad\quad a = 1.06 \text{ m s}^{-2}$

4 (a)

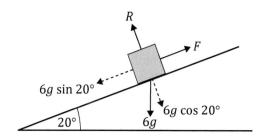

Resolving at right angles to the plane, we have

$$R = 6g \cos 20° = 6 \times 9.8 \times \cos 20° = 55.25 \text{ N}$$

For limiting friction, $F = 6g \sin 20° = 20.11$ N

On the point of slipping, limiting friction $F = \mu R$

Hence least value of μ is $\mu = \dfrac{F}{R} = \dfrac{20.11}{55.25} = 0.36$ (to 2 significant figures)

(b) For motion down the plane, $ma = 6g \sin 20° - F$

so $\quad\quad\quad\quad\quad\quad 6a = 20.11 - (0.2 \times 6g \cos 20°)$

$\quad\quad\quad\quad\quad\quad\quad\quad 6a = 20.11 - (0.2 \times 55.25)$

so $\quad\quad\quad\quad\quad\quad\quad a = 1.51 \text{ m s}^{-2}$

5 (a) First draw a diagram showing all the forces acting on the mass.

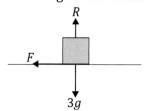

Resolving vertically, we obtain $R = 3g$

$$F = \mu R$$
$$= \frac{6}{49} \times 3 \times 9.8 = 3.6 \text{ N}$$

This frictional force causes a deceleration.

According to Newton's 2nd law of motion, $F = ma$

Hence $\quad -3.6 = 3a$

$\quad\quad\quad\quad a = -1.2 \text{ m s}^{-2}$

BOOST

Grade ⇧⇧⇧⇧

Always check to see if the question asks for the answer to be given to a certain number of significant figures or decimal places. Failure to do this may cost you marks.

In the above diagram we are assuming that the particle is moving to the right so we can take this direction as positive. As the frictional force acts to the left, it will be −3.6 N.

(b) Using $\qquad v^2 = u^2 + 2as$

$$0^2 = 9^2 + 2 \times (-1.2)s$$

Hence, $\qquad s = 33.75\,\text{m}$

6 (a)

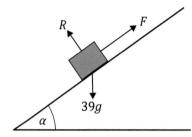

Resolving perpendicular to the plane, we obtain:

$$R = 39g \cos \alpha$$

$$= 39 \times 9.8 \times \frac{12}{13}$$

$$= 352.8\,\text{N}$$

$$F = \mu R$$

$$= 0.3 \times 352.8$$

$$= 105.84\,\text{N}$$

Applying Newton's 2nd law of motion down the slope, we obtain:

$$39g \sin \alpha - F = 39a$$

$$39 \times 9.8 \times \frac{5}{13} - 105.84 = 39a$$

$$a = 1.0554$$

$$= 1.06\,\text{m s}^{-2}$$

(b) Applying Newton's 2nd law of motion up the slope, we obtain:

$$T - 39g \sin \alpha - F = 39a$$

$$T = 147 + 105.84 + 39 \times 0.4$$

$$T = 268.44\,\text{N}$$

As $\tan \alpha = \frac{5}{12}$, the hypotenuse of a right-angled triangle will be of length 13 so $\cos \alpha = \frac{12}{13}$.

Here we are taking down the slope as the positive direction.

Topic 8

1 (a) The weight is the only force acting on the fish (i.e. no air resistance acts). The motion is coplanar (i.e. motion is in a single plane).

(b) Considering the vertical motion, we have

$u = 0\,\text{m s}^{-1}$ (this is the velocity in the vertical direction)

$a = g = 9.8\,\text{m s}^{-2}$

$s = 10\,\text{m}$ (i.e. the height of the eagle)

$t = ?$ (this is the required time of flight)

We can use the following equation of motion:

$$s = ut + \frac{1}{2}at^2$$

$$10 = 0 + \frac{1}{2} \times 9.8t^2$$

Solving gives $t = 1.4285 = 1.43\,s$ (2 d.p.)

(c) Considering the horizontal motion

$$\text{Range} = \text{velocity} \times \text{time of flight}$$

$$= 6 \times 1.4285$$

$$= 8.57\,\text{m}\ (2\ \text{d.p.})$$

2 The horizontal component, $u_x = U\cos\alpha = 30\cos 30° = 25.9808\text{ m s}^{-1}$

The vertical component, $u_y = U\sin\alpha = 30\sin 30° = 15\text{ m s}^{-1}$

Considering the vertical component to find the time of flight and taking the upward direction as positive we have:

$$s = ut + \frac{1}{2}at^2$$

$$s = 15t - \frac{1}{2}gt^2$$

As $s = 0$, $0 = 15t - \frac{1}{2} \times 9.8t^2$

Hence $t\left(15 - \frac{1}{2} \times 9.8t\right) = 0$, so either $t = 0$ or $15 - \frac{1}{2} \times 9.8t = 0$.

The $t = 0$, is when the particle is first projected, so we use the other solution.

Hence, time of flight, $t = 3.06\text{ s}$

Considering the horizontal component to find the range

Range = horizontal velocity × time of flight

$= 25.9808 \times 3.06 = 79.5\text{ m}$

The range (i.e. 79.5 m) is greater than the width of the canyon (i.e. 75 m) so he will clear the canyon.

3 (a)

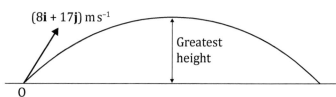

$(8\mathbf{i} + 17\mathbf{j})\text{ m s}^{-1}$ Greatest height   0

Using the equation of motion, $s = ut + \frac{1}{2}at^2$ we can write the following equation in terms of the vectors:

$$\mathbf{s} = \mathbf{u}t + \frac{1}{2}\mathbf{a}t^2$$

$$\mathbf{s} = (8\mathbf{i} + 17\mathbf{j})t - \frac{1}{2} \times 9.81\mathbf{j} \times t^2$$

$$\mathbf{s} = 8t\mathbf{i} + 17t\mathbf{j} - 4.9t^2\mathbf{j}$$

$$\mathbf{s} = 8t\mathbf{i} + (17t - 4.9t^2)\mathbf{j}$$

As this displacement vector is relative to the origin O, this will be the same as the position vector.

Hence position vector relative to O is $\mathbf{r} = 8t\mathbf{i} + (17t - 4.9t^2)\mathbf{j}$

(b) The greatest height occurs when the vertical velocity is zero.

Using $v^2 = u^2 + 2as$ we have

$0^2 = 17^2 - 2 \times 9.8s$

Hence, greatest height = 14.7 m

4 (a) The horizontal component, $u_x = U\cos\alpha = 24.5\cos 30° = 12.25\sqrt{3}\text{ m s}^{-1}$

The vertical component, $u_y = U\sin\alpha = 24.5\sin 30° = 12.25\text{ m s}^{-1}$

Considering the vertical component to find the time of flight and taking the upward direction as positive we have:

$$s = ut + \frac{1}{2}at^2$$

$$s = 12.25t - \frac{1}{2}gt^2$$

As $s = 0$, $\quad 0 = 12.25t - \frac{1}{2} \times 9.8t^2$

Hence $t\left(12.25 - \frac{1}{2} \times 9.8t\right) = 0$, so either $t = 0$ or $12.25 - \frac{1}{2} \times 9.8t = 0$.

Note that g is negative as it acts downwards.

Note that g is negative as it acts downwards.

The $t = 0$, is when the particle is first projected, so we use the other solution.

Hence, time of flight, $t = 2.5$ s

Considering the horizontal component to find the range

$$\text{Range} = \text{horizontal velocity} \times \text{time of flight}$$
$$= 12.25\sqrt{3} \times 2.5 = 53.04 \text{ m}$$

> Note that at the greatest height the vertical component of the velocity is zero.

(b) Considering the vertical motion with the upward direction as positive, we have:

$$v^2 = u^2 + 2as$$
$$0^2 = 12.25^2 - 2 \times 9.8s$$

Hence, the maximum height = 7.66 m

> The negative tells us that the direction of this velocity is opposite to the direction we have taken as positive. Hence this velocity is vertically down.

(c) The horizontal velocity stays constant at $12.25\sqrt{3}$ m s^{-1}

The vertical velocity when it hits the ground can be found using

$$v = u + at$$
$$v = 12.25 - 9.8 \times 2.5 = -12.25 \text{ m s}^{-1}$$

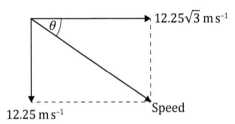

By Pythagoras theorem, $\text{speed} = \sqrt{(12.25\sqrt{3})^2 + (-12.25)^2}$
$$= 24.5 \text{ m s}^{-1} \text{ (downwards)}$$

Angle $\theta = \tan^{-1}\left(\dfrac{12.25}{12.25\sqrt{3}}\right) = 30°$ to the horizontal

5 (a) Horizontal component $= V\cos\alpha = V \times \dfrac{4}{5} = 0.8V$

> $\tan\alpha = \frac{3}{4}$ so if we draw a right-angled triangle with opposite 3 and adjacent 4 then the hypotenuse will be 5 (either use Pythagoras' theorem or use the fact that it is a 3:4:5 triangle).

Vertical component $= V\sin\alpha = V \times \dfrac{3}{5} = 0.6V$

(b) Considering the horizontal motion. The horizontal velocity stays constant. Hence we have, $0.8V \times T = 12$

So $VT = 15$

(c) If you look at the equation $VT = 15$ you can see it contains the time of flight T. We can use the vertical component of the velocity to find the time of flight as follows. We will take the upward direction as positive.

$$s = ut + \frac{1}{2}at^2$$

> Note that g is negative as it acts downwards. Notice that s, the vertical displacement is below the point of projection and is therefore negative.

$$-5.4 = 0.6VT - \frac{1}{2} \times 9.8 \times T^2$$

$$-5.4 = 0.6 \times 15 - 4.9T^2$$

$$4.9T^2 = 14.4$$

Giving $T = \dfrac{12}{7}$ s

Now as $VT = 15$ we have $\dfrac{12}{7}V = 15$ giving $V = 8.75$ m s^{-1}

(d) The horizontal velocity stays constant at $0.8V = 0.8 \times 8.75 = 7$ m s^{-1}

For the vertical motion we use $v = u + at$

$$v = 0.6 \times 8.75 - 9.8 \times \frac{12}{7} = -11.55 \text{ m s}^{-1}$$

By Pythagoras' theorem, $\text{speed} = \sqrt{7^2 + (-11.55)^2} = 13.5$ m s^{-1}

Topic 9

1 (a) $\dfrac{dV}{dt} = -kV$

(b) Separating variables and integrating

$$\int \frac{1}{V} dV = -k \int dt$$

$$\ln V = -kt + C \qquad (1)$$

When $t = 0$, $V = 10\,000$

Substitution of these values in (1) gives

$$\ln 10\,000 = -k(0) + C = C$$

Substitute for C in (1)

$$\ln V = -kt + \ln 10\,000$$

$$\ln V - \ln 10\,000 = -kt$$

$$\ln \frac{V}{10\,000} = -kt$$

Taking exponentials of both sides $\quad \dfrac{V}{10\,000} = e^{-kt}$

$$\therefore V = 10\,000 e^{-kt}$$

(c) (i) When $t = 48$, $V = 4000$

Hence $\quad 4000 = 10\,000 e^{-48k}$

So $\quad e^{-48k} = 0.4$

Taking ln of both sides

$$-48k = \ln 0.4$$

$$k = 0.019$$

Hence $\quad V = 10\,000 e^{-0.019t}$

When $t = 12 \quad V = 10\,000 e^{-0.019 \times 12}$

$$= £7961 \text{ (correct to the nearest pound)}$$

(ii) $V = 10\,000 e^{-kt}$

When $V = 3000$, $3000 = 10\,000 e^{-0.019t}$

So $\quad 0.3 = e^{-0.019t}$

Taking ln of both sides

$$\ln 0.3 = -0.019t$$

Giving $t = 63$ months (to the nearest month)

2 (a) $\dfrac{dC}{dt} = kC$

(b) $\dfrac{dC}{dt} = kC$

Separating the variables and integrating we obtain

$$\int \frac{1}{C} dC = k \int dt$$

$$\ln C = kt + A \qquad (1)$$

Subtract these two
equations to eliminate A.

When $t = 0$, $C = 0.90$
When $t = 2$, $C = 8$
Substitute these values in (1)

$$\ln 0.90 = A$$
$$\ln 8 = 2k + A$$

Then $\quad \ln 0.90 - \ln 8 = A - 2k - A$
giving $\quad \ln 0.90 - \ln 8 = -2k$

$$\ln \frac{0.90}{8} = -2k$$

$$-2.18 = -2k$$

$k = 1.09$ (correct to 3 significant figures)
Substitute for A and k in (1)

$$\ln C = 1.09t + \ln 0.90$$
$$\ln C - \ln 0.90 = 1.09t$$

Taking exponentials of both sides

$$\frac{C}{0.90} = e^{1.09t}$$

$$\therefore C = 0.90e^{1.09t}$$

In the previous examples, a feature of the solutions was that they all involved exponentials, hence the names for such problems are exponential growth or exponential decay. The exponential function occurred in the solutions of differential equations such as

$$\frac{dC}{dt} = kC$$

$$\frac{dV}{dt} = -kV$$

because the right-hand sides involved C^1 or V^1 (i.e. to the power one). Exponentials will not arise when the powers are not unity (i.e. one) such as

$$\frac{dP}{dt} = kP^2 \quad \text{and} \quad \frac{dV}{dt} = -kV^3$$

3 (a)

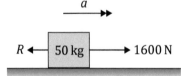

Applying Newton's 2nd law of motion, we obtain:
$$50a = 1600 - R$$
Now $R \propto t$ so $R = kt$
Hence, $50a = 1600 - kt$
When $t = 2$, $a = -4$ so $-200 = 1600 - 2k$
Hence, $k = 900$ and $50a = 1600 - 900t$

$$a = \frac{dv}{dt}$$

Divide both sides of the
equation by 50.

so $\qquad 50\frac{dv}{dt} = 1600 - 900t$

Hence $\qquad \frac{dv}{dt} = 32 - 18t$

(b) Separating variables and integrating we obtain:
$$\int dv = \int (32 - 18t)\,dt$$
$$v = 32t - 9t^2 + c$$
When $t = 2$, $v = 41$
$$41 = 32(2) - 9(2)^2 + c \text{ giving } c = 13$$
Hence $\qquad v = 32t - 9t^2 + 13$
When $v = 28$, $\quad 28 = 32t - 9t^2 + 13$
$$9t^2 - 32t + 15 = 0$$
$$(9t - 5)(t - 3) = 0$$
$$t = \frac{5}{9}, 3 \text{ seconds}$$

> Notice that you are asked to find the times in the question so you are looking for more than one time.

4 (a) $\dfrac{dP}{dt} = kP$

Separating variables and integrating, we obtain:
$$\int \frac{dP}{P} = \int k\,dt$$
$$\ln P = kt + C$$
When $t = 0$, $P = 10$
$$\ln 10 = 0 + C$$
$$C = \ln 10$$

Hence, $\qquad \ln \dfrac{P}{10} = kt$

Taking exponentials of both sides we obtain:
$$\frac{P}{10} = e^{kt}$$
$$P = 10e^{kt}$$

(b) When $t = 1$, $P = 20$ so substituting into $\ln \dfrac{P}{10} = kt$ gives $k = \ln 2$.
$$t = \frac{1}{k} \ln \frac{P}{10}$$
$$t = \frac{1}{\ln 2} \ln (0.1P)$$

When $P = 1\,000\,000$, $\quad t = \dfrac{1}{\ln 2} \ln (0.1 \times 1\,000\,000)$

$t = 16.61$ hours